Routledge Studies in New Media and Cyberculture

1 **Cyberpop**
Digital Lifestyles and
Commodity Culture
Sidney Eve Matrix

2 **The Internet in China**
Cyberspace and Civil Society
Zixue Tai

3 **Racing Cyberculture**
Minoritarian Art and Cultural
Politics on the Internet
Christopher L. McGahan

4 **Decoding Liberation**
The Promise of Free and
Open Source Software
*Samir Chopra and Scott
D. Dexter*

5 **Gaming Cultures and
Place in Asia-Pacific**
*Edited by Larissa Hjorth
and Dean Chan*

6 **Virtual English**
Queer Internets and Digital
Creolization
Jillana B. Enteen

7 **Disability and New Media**
Katie Ellis and Mike Kent

8 **Creating Second Lives**
Community, Identity and
Spatiality as Constructions
of the Virtual
*Edited by Astrid Ensslin
and Eben Muse*

9 **Mobile Technology and Place**
*Edited by Gerard Goggin
and Rowan Wilken*

10 **Wordplay and the Discourse
of Video Games**
Analyzing Words, Design, and Play
Christopher A. Paul

11 **Latin American Identity in
Online Cultural Production**
Claire Taylor and Thea Pitman

12 **Mobile Media Practices,
Presence and Politics**
The Challenge of Being
Seamlessly Mobile
*Edited by Kathleen M.
Cumiskey and Larissa Hjorth*

13 **The Public Space of Social
Media**
Connected Cultures of the
Network Society
Thérèse F. Tierney

14 **Researching Virtual Worlds**
Methodologies for Studying
Emergent Practices
*Edited by Ursula Plesner
and Louise Phillips*

15 **Digital Gaming Re-imagines
the Middle Ages**
Edited by Daniel T. Kline

16 **Social Media, Social Genres**
Making Sense of the Ordinary
Stine Lomborg

Theories of the Mobile Internet

This volume proposes the mobile Internet is best understood
technical "assemblage" of objects, practices, symbolic represent;
riences and affects. Authors from a variety of disciplines disc
mediated through mobile communication, including current
tablet devices. The converging concepts of Materialities (rang)
political economy of communication to physical devices) and
(including cultural values, desires and perceptions) are touchstc
of the chapters in the book.

Andrew Herman is an associate professor of Communication S
Faculty of Arts at Wilfrid Laurier University in Waterloo, Car
written widely in the field of social theory, media and culture
many publications are *The "Better Angels" of Capitalism: Rhe*
tive and Moral Identity Among Men of the American Upper
and his edited collections, *Mapping the Beat: Popular Music an*
rary Cultural Theory (1997) and *The World Wide Web and C*
Cultural Theory (2000).

Jan Hadlaw is an associate professor at York University in Toro;
Her scholarly interests include media history, science and tech
ies, and the role of design in everyday life. Her current resear
the role played by technology and design in the construction
national identity. Her work has appeared in *Space and Cul*
Issues, and *Objets et communication, MEI* (*Médiation et infor*

Thom Swiss is professor of Culture and Teaching at the U
Minnesota. Author of two books of poems, *Measure* and *R*
is the editor of books on popular music, including *Bob Dyl*
61 Revisited (2009), as well as books on new media literature
including *New Media Poetics: Contexts, Technotexts, and The*

17 **The Culture of Digital Fighting Games**
Performances and Practice
Todd Harper

18 **Cyberactivism on the Participatory Web**
Edited by Martha McCaughey

19 **Policy and Marketing Strategies for Digital Media**
Edited by Yu-li Liu and Robert G. Picard

20 **Place and Politics in Latin American Digital Culture**
Location and Latin American Net Art
Claire Taylor

21 **Online Games, Social Narratives**
Esther MacCallum-Stewart

22 **Locative Media**
Edited by Rowan Wilken and Gerard Goggin

23 **Online Evaluation of Creativity and the Arts**
Edited by Hiesun Cecilia Suhr

24 **Theories of the Mobile Internet**
Materialities and Imaginaries
Edited by Andrew Herman, Jan Hadlaw and Thom Swiss

Theories of the Mobile Internet

Materialities and Imaginaries

**Edited by Andrew Herman,
Jan Hadlaw and Thom Swiss**

Routledge
Taylor & Francis Group

NEW YORK AND LONDON

First published 2015
by Routledge
711 Third Avenue, New York, NY 10017

and by Routledge
2 Park Square, Milton Park, Abingdon, Oxon OX14 4RN

*Routledge is an imprint of the Taylor & Francis Group,
an informa business*

Library of Congress Cataloging-in-Publication Data
Theories of the mobile internet : materialities and imaginaries / edited by
    Andrew Herman, Jan Hadlaw, Thom Swiss.
        pages cm — (Routledge studies in new media and cyberculture ; 24)
    Includes bibliographical references and index.
  1. Wireless Internet—Social aspects.   2. Mobile computing—Social
aspects.   I. Herman, Andrew, 1957–   II. Hadlaw, Jan.   III. Swiss,
Thomas, 1952–
    HM851.T498 2014
    302.23'1—dc23
    2014023929

ISBN: 978-0-415-73100-3 (hbk)
ISBN: 978-1-315-84985-0 (ebk)

Typeset in Sabon
by Apex CoVantage, LLC

Contents

List of Figures   ix
Acknowledgements   xi

Introduction: Theories of the Mobile Internet: Mobilities,
Assemblages, Materialities and Imaginaries   1
ANDREW HERMAN, JAN HADLAW AND THOM SWISS

PART I
The Politics of Mobility and Immobility

1   "We Shall Not Be Moved": On the Politics of Immobility   15
    DARIN BARNEY

2   Openness and Enclosure in Mobile Internet Architecture   25
    ALISON POWELL

3   The Materiality of Locative Media: On the Invisible
    Infrastructure of Mobile Networks   45
    JASON FARMAN

4   Labors of Mobility: Communicative Capitalism
    and the Smartphone Cybertariat   60
    ENDA BROPHY AND GREIG DE PEUTER

PART II
Mobile Pasts and Futures

5   Wireless Pasts and Wired Futures   87
    GHISLAIN THIBAULT

6　The Rise, Fall and Future of BlackBerry™ Capitalism　　　105
ANDREW HERMAN AND VINCENT MANZEROLLE

7　Mobile Web 2.0: New Imaginaries of Mobile Internet　　　134
GERARD GOGGIN

8　Future Archaeology: Re-animating Innovation in the
Mobile Telecoms Industry　　　149
LAURA WATTS

PART III
Living Mobile Lives

9　The Mobile Phone (and Texts) as a Taken-for-Granted
Mediation　　　171
RICH LING

10　New and Old, Young and Old: Aging the Mobile
Imaginary　　　187
BARBARA CROW AND KIM SAWCHUK

11　"I'm Melvin, a 4G Hot Spot"　　　200
THOM SWISS

12　A Hole in the Hand: Assemblages of Attention　　　212
and Mobile Screens
J. MACGREGOR WISE

13　Apps and Drive　　　232
JODI DEAN

Contributors　　　249
Index　　　253

List of Figures

3.1 The door that led to the Equinix Internet Peering Point
on the outskirts of Washington, DC. Image © 2011
Jason Farman.   46

3.2 The inside of the Equinix Internet Peering Point
in Ashburn, Virginia. © 2011 Equinix, Inc.   47

3.3 The base of the cell tower that my phone connected
with near the Equinix Peering Point, showing the wired
infrastructure that my data flowed through. Image © 2013
Jason Farman.   48

3.4 An image of my check-in to Equinix on Foursquare,
which broadcasted my location, activity and companions
to those in my network. Image © 2011 Jason Farman.   50

3.5 A cell tower disguised as an evergreen tree in Silver Spring,
Maryland. Image © 2012 Jason Farman.   56

7.1 "Mobile Internet with the Best Coverage . . . The Internet
Where All is Possible" (Claro advertising billboard, Cusco,
Peru, 2011).   141

7.2 "With an Android Tigo, You're in Everything" (Tigo
advertising billboard, Cartegna, Columbia, 2011).   142

7.3 "Shared, Life is More" (Movistar advertising billboard,
Lima, Peru, 2011).   143

9.1 Ratio of texts to calls (three-period moving average
superimposed on raw data) based on 13,931 users in
the Telenor net in Norway in 2009.   179

12.1 Screenshot of the app LiveCams.   219

12.2 Japan.   220

12.3 USA.   221

12.4 Sweden.   221

12.5 Russia.   222

Acknowledgements

This book has its origins in a workshop funded the Social Science and Humanities Research Council of Canada. Funding for the workshop was also provided by the Communication Studies Department, the Cultural Analysis and Social Theory Graduate Program and the Offices of the Provost and Vice-President-Academic, Vice-President-Research and the Dean-Faulty of Arts, all at Wilfrid Laurier University. We'd like to thank Cheryl Dipede and Christopher Moorehead, graduate students in the Masters Program in Design at York University, for their assistance organizing workshop publicity and participant travel. Material support was provided by Communitech and the Canadian Digital Media Network. We would also like to especially thank Harry Froklage, associate director of the Office of Development and Alumni Relations and Wilfrid Laurier University, for his indefatigable efforts in raising money for the workshop.

Subsequent support for the Materialities and Imaginaries of the Mobile Internet Project was provided by the Graphics, Animation and New Media-Network Centre of Excellence (GRAND-NCE), funded by the Government of Canada Network Centres of Excellence program, as well as the Wilfrid Laurier University Office of the Vice-President Research. The first editor would also like to personally thank Abby Goodrum, Laurier's erstwhile Vice-President Research, for her support of this book project and her invitation to the join the GRAND-NCE. Many thanks are due to Anne Ramsey and Leigh Robinson, who worked as "HQP" ("Highly Qualified Personnel") on the MIMI and GRAND-NCE projects.

Andrew Herman would also like to thank Rosemary J. Coombe for her love and her exemplary devotion to the vocation of critical scholarship. Jan Hadlaw thanks Ben Bradley as a matter-of-fact. Thom Swiss thanks his colleagues at the University of Minnesota and Cynthia Lewis for her many deft moves.

Introduction

Theories of the Mobile Internet: Mobilities, Assemblages, Materialities and Imaginaries

Andrew Herman, Jan Hadlaw and Thom Swiss

As the Internet enters its third decade as a popular medium, one of the most important aspects of its evolution has been the transformation of the digital way of life from a state of being "wired" to that of being "wireless" and "mobile."[1] While the *wireless* Internet refers to the increasing diffusion of "Wi-Fi" connectivity (and its more powerful successors), the *mobile* Internet entails wireless connectivity and more. The technological advancements in platforms for the latter are driving the expansion of the provision of the former.

The mobile Internet is manifested in the proliferation of practices mediated through "3G" (third generation) and IMT-Advanced (commonly called "4G" or fourth generation) data transfer networks that are accessed through mobile communication devices ranging from smartphones (such as the iPhone or the many Android OS–based devices) to a wide array of tablet devices (such as the iPad) through which Internet experiences truly become portable across a multiplicity of spaces and places of everyday life.[2] Until the development of these devices, mobile communication was more or less comprised of untethered telephony with the added (and originally unintended) benefit of SMS texting (Goggin 2006). As we know, 3G and 4G devices embody the convergence of new multimedia forms, applications and platforms that extend the possibilities of mobile communication far beyond talking and texting. Indeed mobile Internet users—particularly those who accessing video and gaming content—are increasingly consuming the vast majority of wireless bandwidth.

While there has been a tremendous flowering of scholarly work on mobile telephony and communication over the past five years, Internet scholars have been slower to consider the growing mobility of the Internet.[3] This *aporia* in critical Internet studies is problematic for, as Mimi Ito has argued, the development of the mobile Internet "demands a set of engagements at various methodological and theoretical points that differ substantially from Internet study" (2005, 5). Indeed, from our perspective, making sense of the mobile Internet necessitates the development of a unique analytic that closes the gap between Internet scholarship on the one hand and scholarship on mobile communication on the other.

MOBILITIES AND ASSEMBLAGES

This book takes up the theoretical and conceptual challenge that Ito posited—but did not elaborate—by proposing an approach that focuses on how the mobile Internet shapes and is shaped by the emergence of what Mimi Sheller and John Urry term the "new *mobilities* paradigm" (2007, emphasis added). This paradigm emphasizes multiple and linked registers of mobility which include (but are not limited to) the physical movement of travel and migration, the symbolic movement of mediated communication and the circulation of cultures, and the virtual movement in and through digitized spaces (Urry 2007; Elliott and Urry 2010, Cresswell 2010). All three registers, Sheller and Urry argue, converge in the practices of mobile telephony which enable the capacity of "interacting and communicating on the move, of being in a sense present while apparently absent" within the same spatio-temporal coordinates (2007, 207). Accordingly, this volume seeks to address two apposite issues: First, how do mobile communications and the mobile Internet create zones of connectivity that are fluid, transportable, and meaningful? Second, what are the structures of *immobility* (i.e., server farms, wireless towers, Wi-Fi hotspots, fiber optics of Internet backbones, and so on) that also serve to constitute the mobile Internet?

To address these questions, authors in this book construct the subject of the mobile Internet as an *assemblage* of multidimensional socio-technical practices in which the materialities and imaginaries of manifold technologies of mobile communications converge. At its most abstract level in terms of social theory, "assemblage thinking allows us to foreground ongoing processes of composition across and through different human and non-human actants," a perspective that is absolutely crucial to understanding the mobile Internet as a contingent totality (Anderson et al. 2012, 172). Digital gaming scholar T. L. Taylor describes the concept of the assemblage—derived from Gilles Deleuze and Félix Guattari (1987)—as an "alternative heuristic" for understanding complex socio-technical systems: "the notion of assemblage is one way to help us understand the range of actors (system, technologies, player, body, community, company, legal structures, etc.), concepts, practices, and relations" (Taylor 2009, 332). Taylor is noting here the heterogeneity of the assemblage as a network of imaginary and material elements, elements that are not randomly connected but are structured with a particular vector of force and effectivity in time and space. Jennifer Slack and the author of a final chapter in this volume, Macgregor Wise (2005), render this heuristic more nuanced by distinguishing between "articulation" and "assemblage." A socio-technological articulation refers to how an elemental ensemble of objects, social practices, symbolic representations, experiences and affects are drawn together in a specific and contingent unity. An assemblage is a socio-technical articulation that is, to use a Deleuzian metaphor, "constellated" into a "particular dynamic form" that envelops and gives form to spatio-temporal territories (Slack and Wise 2005, 128).

From our perspective, the concept of assemblages is as important methodologically as it is ontologically. John Law, in *After Method*, emphasizes the *ad hoc* and active meaning of the concept. An assemblage is not a fixed arrangement or state of affairs, but rather an open-ended, "uncertain and hesitant process" (2004, 41). Assemblages are both objects and activities. The mobile Internet is neither one nor the other, but rather "the event . . . of their reciprocal configuration" (Taylor 2009, 332). The notion of the assemblage makes understanding the mobile Internet *in situ*—whether that situation be physical, symbolic or virtual—possible. The materialities and imaginaries of the mobile Internet are fluid and dynamic, and thus their apprehension and analysis must be likewise methodologically open-ended and plural.

MATERIALITIES AND IMAGINARIES

The assemblages of the mobile Internet are comprised of articulations of materialities and imaginaries. The term "materialities" invokes two traditions of materialist analysis, both of which have long, storied and influential purchase in communication/cultural/media studies. The first derives from the neo- and autonomist Marxist materialism of the political economy of communication (Mosco 1996; Dyer-Witheford 1999; Dyer-Witheford and de Peuter 2009). From this perspective, the centrality of the mobile Internet in constructing a distinctive form of digital and/or informational capitalism is paramount. If, as Benedict Anderson (2006) famously argued, the apparatus of print technologies ushered in the era of "print capitalism," is it possible or fruitful to speak of the emergence of "smartphone" or "notepad" capitalism? What are the arrays of social relations of power and exploitation—embodied in markets and their constituent social practices—that characterize this new form of capitalism as a distinctive social formation? What are the "circuits of interactivity"—between smartphone and user, between telecoms and regulatory agencies, between applications and software designers, between Google or Apple and regimes of intellectual property and so on—that comprise the "empire" of network informational capitalism (Kline, Dyer-Witheford and de Peuter 2003)?

The second tradition derives from the materialism rooted in the well-known "medium theory" of Innis and McLuhan (Meyrowitz 1994; Deibert 1997), which is in turn a major inspiration for the contemporary "media materialism" of Friedrich Kittler (1999; Gane 2005; Parikka 2012; Pels et al. 2002). This evolving and disparate body of work takes seriously the idea that the material forms of mobile communication—from the digital architecture of 3G networks to the design of the devices themselves—are just as important as the meaning of the messages they convey and carry. How do the materialities of the mobile Internet manifest themselves in the everyday encounters of users and their devices? To what extent does the design of haptic interfaces of mobile device construct or determine the mobile experience? Concern with such materialities may also entail a concern with their corporealities. Pushed to the limit, the materialities of the mobile Internet encompass the connectivities that the ITU has called the emergent "Internet of Things" (2005), where H2H (human-to-human) communication is increasingly supplemented by H2T (human-to-things) and T2T (things-to-things) (Varnelis and Friedberg 2007, 37). The important of the emergence of the Internet of Things to issues of mobility is underscored by the fact that wireless bandwidth usage will be increasingly dominated by T2T communication (Lawson 2014).

All technologies—in their development, dissemination and usage—are embedded within and animated by "social imaginaries." According to Charles Taylor (2004), a social imaginary is an epistemological and ontological framework of cultural value and identity that is at once more supple and yet firmly embedded in quotidian perceptions and practices than ideologies *per se*. As Andrew Herman (2010 has argued, "Social imaginaries then are not simply a set of *ideas* about the social world: they are pragmatic templates for social *practice*. In this way, social imaginaries operate as forms of power-knowledge, enabling some social actions and constraining others as they provide a map of the social as moral space that is delineated along existential, normative and utopian dimensions" (190).

Patrice Flichy (2007) observed in *The Internet Imaginaire* that there are few technological developments in modernity that have not been animated by such an imaginary. Among the key questions the chapters in this collection consider are how the mobile Internet is constructed through such imaginaries and how these imaginaries privilege particular definitions of personal being and valorize specific desires of social becoming. As Jan Hadlaw (2011) more recently reminds us, new communications technologies have historically acted as the sites on which negotiations over—often competing—social values and utopian desires have been played out. The utopian promise of technological progress associated with the mobile Internet continues this tradition. What do desires to project ourselves into the future tell us about our social and technological present? What do such social and technological imaginaries tells us about past imaginaries where other wireless utopias were projected onto the future, and how can bringing an historical perspective to understanding the imaginaries of the mobile Internet provide us with insight into contemporary imaginaries (Gitelman 2006)?

THE PLAN OF THE BOOK

Following Slack and Wise's analytic of articulations and assemblage, the book will explore three distinct *topoi* of the mobile Internet and its spatio-temporal territorialization(s): the politics of mobility and its other; past and future conceptions of Internet mobilities; and mobilities of the Internet in everyday life.

Part I: The Politics of Mobility and Immobility

As we noted earlier, mobile communications and the mobile Internet create zones of connectivity that are fluid, transportable and meaningful, but
the structures and politics of *immobility* also serve to constitute the mobile
Internet. Involuntary immobility enforced on those who occupy the lower
registers of various socio-economic hierarchies is, of course, inegalitarian
and unjust. But what about immobilities enforced *by* those who occupy
these lower registers upon those who would prefer that they, and things,
just *keep moving*? This section begins with Darin Barney's "We Shall Not Be
Moved," in which he interrogates the moral economy of mobility in which
freedom is assumed to be contingent upon open access to mobile communication. Barney argues that, to the contrary, mobility is not *ipso facto* a
ground of freedom and empowerment; indeed a politics of *immobility* in
relation to work and labor may be the progressive line of flight. Alison Powell's "Openness and Enclosure in Mobile Internet Architecture" takes issue
with the always-already necessary convergence of "open" architecture and
"open" digital public sphere that has dominated liberatory rhetoric in digital culture since the 1970s. She examines how the materialities and imaginaries of wired and wireless Internet embodies this conflation of openness
with freedom, with special eye toward the development of the hardware
of mobile Internet devices. In "The Materiality of Locative Media," Jason
Farman offers a thoughtful mediation on the relationship between space as
location, place as identity, and how both are performed through the unseen
layers of the material infrastructures of mobile networks. Farman's ethnographic account of a visit to one the "doors to the Internet" in the United
States—the Equinix Data Center outside of Washington, DC—becomes the
foundation for the development of what he terms an "object-oriented phenomenology" that will, hopefully, render the "vibrant matter" of the material infrastructures of the mobile Internet visible. Part I concludes with Enda
Brophy and Greig de Peuter's chapter on "Labors of Mobility." They map
the value chain of "communicative capitalism" by tracing the flows of labor
that produce the "smartphone" at multiple registers of materiality. In so
doing, they address what they term the "labor blindspot" of contemporary
critical analysis of the mobile communication and the mobile Internet.

Part II: Mobile Pasts and Futures

As a heterogeneous assemblage of materialities and imaginaries, the mobile
Internet embodies and instantiates multiple temporalities. Some of these
temporalities cohere synchronically and are structurally present in what
Siegfried Zielinski (2008) terms the "deep time" of contemporary forms and
practices of the Internet gone mobile, such as the smartphone or protocols of
wireless Internet connectivity. Still other temporalities manifest themselves
in diachronic genealogies of the past and present, as well as histories of the

future. This section considers both kinds of temporal modalities, beginning with a look at the futures of wirelessness and mobility as imagined in the past and ending with imagining the future viewed through the optic of stone-age megaliths. In "Wireless Pasts and Wired Futures" Ghislain Thibault examines the technological imaginary of progress and the "dematerialization" of communication infrastructure, an entirely contingent articulation that produced the ideological coupling of wirelessness with mobility, and mobility with freedom, in the late nineteenth century. Andrew Herman and Vincent Manzerolle examine the rise and fall of perhaps the first iconic device of the mobile Internet era—the BlackBerry smartphone—in order to develop a more coherent and durable conceptualization of digital and informational capitalism that has theoretical implications far beyond the life of the device or the company that made it. They argue that the imaginary of a wireless future of mobility and freedom becomes tightly intertwined with the materialities of ubiquitous connectivity that come to dominate the quotidian world of digital labor in the new era of "informational capital/-ism." Gerard Goggin's chapter on "Mobile Web 2.0" closely interrogates two of the conceptualizations of the "imaginary" most influential in the intellectual shaping this project, that of Canadian philosopher Charles Taylor (2004) and Patrice Flichy (2007). Leveraging Taylor's and Flichy's complementary understandings of social imaginaries, Goggin examines multiple histories of mobile Internet assemblages: early wireless Internet protocols; the articulation of the development of mobile handsets with ideologies of ICT4D in the Global South, and the rise of locative media. The combined and uneven development of these imaginaries, argues Goggin, provides a fruitful framework for interrogating the emergent materialities of the mobile Internet as a global phenomenon. Part II concludes with Laura Watts innovative auto-ethnography "Future Archaeology: Re-animating Innovation in the Mobile Telecoms Industry." Reflecting upon her experience in the mobile telecom industry in the early part of the century, Watts shifts the analytical terrain of the territorializations of the mobile Internet away from the historical and genealogical toward the archaeological and the futurological, examining the industry fixation on the archaic figure of the stone-age "pebble," in spite of all revolutionary rhetoric of technological progress to the contrary.

Part III: Living Mobile Lives

The final part of the book considers a range of assemblages of the mobile Internet as they territorialize different quotidian spatialities and temporalities. Rich Ling offers a critical phenomenology of the quotidian experience of the mobile Internet as taken-for-granted lifeworld by examining the complex interplay of practices of social exclusion, social inefficiency and social interaction that is territorialized through the social mediation of the smartphone. In their chapter "New and Old, Young and Old: Aging the Mobile Imaginary," Barbara Crow and Kim Sawchuk consider the articulation of aging, mobility-as-movement and mobile telephony as imagined by the wireless telecom industry as well as social scientific studies of mobile communication. They offer a critical assessment of the tendential force of a mobile assemblage that privileges youth. Thom Swiss troubles the ideological articulation of mobility and freedom through a multi-modal ethnography of homelessness and mobile communication in "I'm Melvin, a 4G Hot Spot." Swiss reflects upon the "conceit" of "assemblage" as an "alternative heuristic that mirrors the fragmented lives of the homeless." As Swiss suggests, the wireless connectivity of one body can be literally dependent upon the displacement and dislocation of another body in complex relations of power and inequity that only a poetic method of methodological connectivity can illuminate. Macgregor Wise's chapter asks us to consider the phenomenological moment of quotidian connectivity that we experience as we gaze at and engage with the mobile devices in our hands. According to Wise, this moment is dis/embodied in the "hole in the hand"—or the screen of our mobile devices—that is a portal and nexus of the "clickable world" of the mobile Internet. Wise reflects upon what he terms the "ethics of attention" in this assemblage that suggests that we inhabit vectors of distraction of view, always-already leading elsewhere but arriving nowhere. Jodi Dean concludes the section—and the book—with a critical account of the rise and hegemony of apps in the constitution of the mobile Internet experience. For Dean, apps are fasteners of attachment and enthrallment. Although one can differ with Dean's argument that apps are the functional *raison d'etre* for smartphones, this does not detract from the fact that apps are "affective machines" that bind us to the supply chains of "drive" that are the sinews of "communicative capitalism."

NOTES

1. According to the OECD, Wireless broadband penetration has grown to 68.4% in the OECD area, according to June 2013 data, meaning there are now more than two wireless subscriptions for every three inhabitants. ("OECD broadband statistics update" January 9, 2014)
2. "3G" refers to a set of technical standards for the mobile communication devices developed by the International Telecommunications Union. According to the ITU, "Third generation (3G) systems promise faster communications services, including voice, fax and Internet, anytime and anywhere with seamless global roaming. ITU's IMT-2000 global standard for 3G has opened the way to enabling innovative applications and services (e.g. multimedia entertainment, infotainment and location-based services, among others)." (www.itu.int/osg/spu/ni/3G/technology/index.html). Interestingly, prior to January 2012, there is no so such thing as "4G" apart from industry hype as the technical standards for IMT-Advanced (aka 4G) had not been agreed upon. Such standards were announced by the ITU on January 18, 2012 (www.itu.int/ITU-D/tech/MobileCommunications/IMT_INTRODUCING/IMT_2G3G4G.html) The technical distinctions between the systems are very important for product differentiation, market strategies and governmental regulation (or lack thereof), especially in the EU and Asia

3. For example, see: Arceneaux and Kavoori 2012; Caron and Caronia 2007; Crow, Longford and Sawchuk 2010; Castells, et. al 2007; de Souza e Silva and Frith 2012; Farman 2011; Farman 2012; Goggin 2006, 2009, 2010; Goggin and Hjorth 2008; Green and Haddon 2009; Hjorth, Burgess and Richardson 2012; Katz and Aakhus 2002; Höflich et al. 2010; Katz 2003, 2006, 2008, 2011; Ling 2004, 2008; Ling and Campbell 2009, 2011; Ling and Donner 2009; Wilken and Goggin 2012.

SOURCES

Anderson, B. *Imagined Communities: Reflections on the Origins and Spread of Nationalism*, 2nd ed. London: Verso, 2006.
Anderson, B., M. Kearnes, C. McFarlane and D. Swanton. "On Assemblages and Geography." *Dialogues in Human Geography*. no. 2 (2012): 171–189.
Arceneaux, N., and A. Kavoori. *The Mobile Media Reader*. New York: Peter Lang, 2012.
Barney, D. *The Network Society*. Malden: Basil Blackwell, 2004.
Benkler, Y. *The Wealth of Networks*. New Haven: Yale University Press, 2006.
Canadian Radio-Television and Tele-Communications Commission. "Communications Monitoring Report 2009." August 5, 2009. Accessed September 5, 2009. www.crtc.gc.ca/eng/publications/reports/PolicyMonitoring/2009/cmr.htm
Carey, J. *Communication as Culture, Revised Edition: Essays on Media and Society*. New York: Routledge, 2008.
Caron, A., and L. Caronia. *Moving Cultures: Mobile Communication in Everyday Life*. Montreal and Kingston: McGill-Queen's University Press, 2007.
Castells, M. *The Rise of The Network Society*. Oxford: Blackwell, 1996.
Castells, M., M. Fernandez-Ardevol, J. Qiu and A. Sey. *Mobile Communication and Society: A Global Perspective*. Cambridge, MA: MIT Press, 2007.
Cresswell, T. "Towards a Politics of Mobility." *Environment and Planning D: Society and Space*. no. 1 (2010): 17–31.
Crow, B., M. Longford and K. Sawchuk (eds.). *The Wireless Spectrum: The Politics, Practices, and Poetics of Mobile Media*. Toronto: University of Toronto Press, 2010.
Cumiskey, K., and L. Hjorth (eds.). *Mobile Media Practices, Presence and Politics: The Challenge of Being Seamlessly Mobile*. New York: Routledge, 2013.
Deibert, R. *Parchment, Printing, and Hypermedia*. New York: Columbia University Press, 1997.
Delueze, G., and F. Guattari. *A Thousand Plateaus: Capitalism and Schizophrenia*. Minneapolis: University of Minnesota Press, 1987.
de Souza e Silva, A., and J. Frith. *Mobile Interfaces in Public Spaces: Locational Privacy, Control, and Urban Sociability*. New York: Routledge, 2012.
Dreyfus, H. *Being-in-the-World: A Commentary on Heidegger's Being and Time*, Division I. Cambridge, MA: MIT Press, 1990.
Dyer-Witheford, N. *Cyber-Marx: Cycles and Circuits of Struggle in High Technology Capitalism*. Urbana: University of Illinois Press, 1999.
Dyer-Witheford, N., and G. de Peuter. *Games of Empire: Global Capitalism and Video Games*. Minneapolis: University of Minnesota Press, 2009.
Elliott, A., and J. Urry. Mobile *Lives*. New York: Routledge, 2010.
Farman, J. *Mobile Interface Theory: Embodied Space and Locative Media*. New York: Routledge, 2011.
Farman, J. (ed). *Digital Storytelling and Mobile Media: Narrative Practices with Locative Technologies*. New York: Routledge, 2012.

Flichy, P. *The Internet Imaginary.* Cambridge, MA: MIT Press, 2007.

Galloway, A., and E. Thacker. *The Exploit: A Theory of Networks.* Minneapolis: University of Minnesota Press, 2007.

Gane, N. "Radical Post-Humanism: Friedrich Kittler and the Primacy of Technology." *Theory, Culture & Society.* no. 3 (2005): 25–41.

Giddings, S. "Playing with Theory: Videogames, the Technological Imaginary and a New Media Studies." Paper presented at the "Playing With The Future: Development and Directions in Computer Gaming" conference, University of Manchester, April 5–7, 2002.

Gitelman, L. *Always Already New: Media, History, and the Data of Culture.* Cambridge, MA: MIT Press, 2006.

Goggin, G. *Cell Phone Culture: Mobile Technology in Everyday Life.* New York: Routledge, 2006.

Goggin, G. *Mobile Phone Cultures.* New York: Routledge, 2009.

Goggin, G. *Global Mobile Media.* New York: Routledge, 2010.

Goggin, G., and L. Hjorth (eds.). *Mobile Technologies: From Telecommunications to Media.* New York: Routledge, 2008.

Green, N., and L. Haddon. *Mobile Communications: An Introduction to New Media.* London: Berg, 2009.

Hadlaw, J. "Saving Time and Annihilating Space: Discourses of Speed in AT&T Advertising, 1909–1929." *Space and Culture.* no. 2 (2011): 85–113.

Hayles, N.K. *How We Became Posthuman: Virtual Bodies in Cybernetics, Literature, and Informatics.* Chicago: University of Chicago Press, 1999.

Haraway, D. *A Cyborg Manifesto: Science, Technology, and Socialist-Feminism in the Late Twentieth Century. Simians, Cyborgs and Women: The Reinvention of Nature.* New York: Routledge, 1991.

Hearn, A. "Meat, Mask, Burden: Probing the Contours of the Branded Self." *Journal of Consumer Culture.* no. 2 (2008): 197–217.

Heiddeger, M. *Poetry, Language and Thought.* New York: Harper Perennial, 2001.

Herman, A. "Manifest Dreams of Connectivity and Communication or, Social Imaginaries of the Wireless Commons." In *The Wireless Spectrum: The Politics, Practices, and Poetics of Mobile Media,* edited by B. Crow, M. Longford and K. Sawchuk, 187–198. Toronto: University of Toronto Press, 2010.

Hjorth, L., J. Burgess and I. Richardson. *Studying Mobile Media: Cultural Technologies, Mobile Communication, and the iPhone.* New York: Routledge, 2012.

Höflich, J.R., G. F. Kircher, C. Linke and I. Schlote. *Mobile Media and the Change of Everyday Life.* Berlin: Peter Lang, 2010.

Horrigan, J. "Wireless Internet Use." Pew Internet & American Life Project. July 22, 2009. Accessed September 5, 2009. www.pewinternet.org/Reports/2009/12-Wireless-Internet-Use.aspx?r = 1

Innis, H. *The Bias of Communication,* 2nd ed. Toronto: University of Toronto Press, 2008.

International Telecommunications Union. *ITU Internet Reports: The Internet of Things.* Geneva: International Telecommunications Union, 2005.

International Telecommunications Union. "What Really Is a Third Generation (3G) Mobile Technology." 2009. Accessed September 6, 2009. www.itu.int/ITU-D/imt-2000/DocumentsIMT2000/What_really_3G.pdf

Ito, M. "Introduction: Personal, Portable, Pedestrian." In *Personal, Portable, Pedestrian: Mobile Phones in Japanese Life,* edited by M. Ito, D. Okabe and M. Matsuda, 1–18. Cambridge, MA: MIT Press, 2005.

Ito, M. Introduction to *Networked Publics,* by K. Varnelis. Cambridge, MA: MIT Press, 2008.

Katz, J. *Machines That Become Us: The Social Context of Personal Communication Technology.* New Brunswick: Transaction Publishers, 2003.

Katz, J. *Magic in the Air: Mobile Communication and the Transformation of Social Life*. New Brunswick: Transaction Publishers, 2006.

Katz, J. *Handbook of Mobile Communication Studies*. Cambridge, MA: MIT Press, 2008.

Katz, J. E. (ed.). *Mobile Communication: Dimensions of Social Policy*. Piscataway: Transaction Publishers, 2011.

Katz, J., and M. Aakhus. *Perpetual Contact: Mobile Communication, Private Talk, Public Performance*. Cambridge, UK: Cambridge University Press, 2002.

Kittler, F. *Gramophone, Film, Typewriter*. Palo Alto: Stanford University Press, 1999.

Kline, S., N. Dyer-Witheford and G. de Peuter. *Digital Play: The Interaction of Technology, Culture and Marketing*. Montreal and Kingston, Canada: McGill-Queens University Press, 2003.

Latour, B. *Reassembling the Social: An Introduction to Actor-Network Theory*. New York: Oxford University Press, 2007.

Law, J. *After Method: Mess in Social Science Research*. New York: Routledge, 2004.

Lawson, Stephen. "Cisco Unveils 'Fog Computing' to Bridge Clouds and the Internet of Things." *PC World*. January 29, 2014. Accessed March 26, 2014. www.pcworld.com/article/2092660/cisco-unveils-fog-computing-to-bridge-clouds-and-the-internet-of-things.html

Ling, R. *The Mobile Connection: The Cell Phone's Impact on Society*. San Francisco: Morgan Kaufman, 2004.

Ling, R. *New Tech, New Ties: How Mobile Communication Is Reshaping Social Cohesion*. Cambridge, MA: MIT Press, 2008.

Ling, R., and S. W. Campbell. *The Reconstruction of Space and Time: Mobile Communication Practices*. New Brunswick: Transaction Publishers, 2009.

Ling, R., and S. W. Campbell. *Mobile Communication: Bringing Us Together and Tearing Us Apart*. Piscataway: Transaction Publishers, 2011.

Ling, R., and J. Donner. *Mobile Phones and Mobile Communication*. Cambridge, UK: Polity, 2009.

Meyrowitz, J. "Medium Theory." In *Communication Theory Today*, edited by D. Crowley and P. Heyer, 50–77. Cambridge, UK: Polity, 1994.

Miller, V. "New Media, Networking and Phatic Culture." *Convergence: The International Journal of Research into New Media Technologies*. no. 4 (2008): 387–400.

Morley, D. *Home Territories: Media, Mobility and Identity*. New York: Routledge, 2004.

Mosco, V. *The Political Economy of Communication*. Newbury Park: Sage, 1996.

Mosco, V. *The Digital Sublime: Myth, Power and Cyberspace*. Cambridge, MA: MIT Press, 2004.

Nye, D. *American Technological Sublime*. Cambridge, MA: MIT Press, 1994.

OECD, "OECD Broadband Statistics Update." January 9, 2014. Accessed March 19, 2014. www.oecd.org/sti/broadband/broadband-statistics-update.htm

Parikka, J. "New Materialism as Media Theory: Medianatures and Dirty Matter." *Communication and Critical/Cultural Studies*. no. 1 (2012): 95–100.

Pels, D., K. Hetherington and F. Vandenberghe. "The Status of the Object: Performances, Mediations, and Techniques." *Theory, Culture & Society*. no. 5/6 (2002): 1–21.

Peters, J. *Speaking into the Air: A History of the Idea of Communication*. Chicago: University of Chicago Press, 2001.

Sheller, M., and J. Urry. "The New Mobilities Paradigm." *Environment and Planning A*. no. 2 (2007): 207–226.

Slack, J. D., and J. M. Wise. *Culture + Technology: A Primer*. Vienna: Peter Lang, 2005.

Taylor, C. *Modern Social Imaginairies*. Durham: Duke University Press, 2004.

Taylor, T. L. "The Assemblage of Play." *Games and Culture.* no. 4 (2009): 331–339.

Terranova, T. *Network Culture: Politics for the Information Age.* London: Pluto Press, 2004.

Urry, J. *Mobilities.* Cambridge, UK: Polity, 2007.

Van Dijk, J. *The Network Society.* Thousand Oaks: Sage, 2006.

Varnelis, K., and A. Friedberg. "Place: The Networking of Public Space" in *Networked Publics,* edited by K. Varnelis, 15–42. Cambridge, MA: MIT Press, 2007.

Wilken, R., and G. Goggin. *Mobile Technology and Place.* New York: Routledge, 2012.

Wittel, A. "Toward a Network Sociality." *Theory, Culture & Society.* no. 6 (2001): 51–57

Zielinski, Siegfried. *Deep Time of the Media: Toward an Archaeology of Hearing and Seeing by Technical Means.* Cambridge, MA: MIT Press, 2008.

Part I

The Politics of Mobility and Immobility

1 "We Shall Not Be Moved"

On the Politics of Immobility

Darin Barney

"I am going nowhere." That's Willie Corduff, a farmer, protesting the proposed landfall of Shell's Corrib Gas Pipeline at Glengad Beach, near Rossport in the West of Ireland. Corduff was jailed for ninety-four days in 2005 for refusing to allow the company to enter his land to lay a high-pressure, raw gas pipeline. Reflecting on his own resistance to the pipeline, which saw him and his son arrested and jailed for obstructing the development, Rossport lobster and crab fisherman Pat O'Donnell observed, "This could go on forever" (Domhnaill 2010). By "this" he meant the political struggle of his community to retain some degree of control over the immediate material conditions of their economic and social lives.

This chapter is not about pipelines or fishers or farmers. Instead, my aim is to ask a few questions about the moral valorization of the mobile Internet and the prevalent cultural designation of mobile access to information and communication networks as something basically "good," something like freedom. More broadly, these are questions about the critical status of the norm of mobility itself, upon which this valorization of mobile and mobilizing technologies at least partly hangs. My concern is with the status of the norm of mobility in relation to the possibility of politics and it is in this respect that the story of the good people of Rossport serves as an instructive introduction. Cultural geographer Timothy Cresswell (2010, 21) defines politics as "the social relations that involve the production and distribution of power," and the politics of mobility as "the ways in which mobilities are both productive of such social relations and produced by them," adding that "speeds, slowness and immobilities are all related in ways that are thoroughly infused with power and its distribution." This is undoubtedly true but, in what follows, I propose a slight shift in emphasis, from politics understood as the *distribution* of power to politics understood as the *disruption* of power and, correspondingly, from the politics of *mobility* to the politics of *immobility*.

Among the six elements Cresswell (2010, 26) lists as comprising the politics of mobility is the question "when and how does it stop?" By this he means to draw attention to the tendency of contemporary injustices to take the form of friction, experienced by those whose mobility is impeded when they might otherwise choose to keep moving. Involuntary immobility enforced *upon* those who occupy the lower registers of various socio-economic hierarchies is certainly one manner in which inegalitarian and unjust distributions of power are currently manifested and maintained. However, I would like to explore the opposite dynamic, whereby immobilities enforced *by* those who occupy these lower registers upon those who would prefer that they, and things, just *keep moving* become a significant source of political disruption. Cresswell's politics of mobility implies that the question of "where and when does it stop?" refers primarily to the mobility of individuals who would like to keep moving but are prevented from doing so by powerful actors and structures over which they have little or no influence. The operative question in this situation becomes: "Is stopping a choice, or is it forced?" (Cresswell 2010, 26). By contrast, from the perspective of a politics of *immobility*, we might instead consider that "where and when does it stop?" is often the question asked by those who (like the people of Rossport) find themselves in a situation where they have no choice but to act, often collectively, to disrupt some force that is moving inexorably against them. Of course, such disruptions often take time. Here, I suggest that in a material context in which mobility and its technologies (including things like gas pipelines and wireless telephone networks) are structurally related to economic power and therefore culturally normalized, the possibility of politics might rely precisely on "going nowhere" and "going on forever."

   The cultural valorization of mobile information and communication technologies (which has now been fairly generalized in commercial advertising, popular culture and economic development discourse) relies on a simultaneous, and perhaps even prior, elevation of mobility itself to the status of an unimpeachable norm, one that corresponds roughly with freedom, and which via this correspondence articulates with liberal discourses of rights, justice and democracy. Cresswell (2010, 21) captures this well when he writes:

> Some of the foundational narratives of modernity have been constructed around the brute fact of moving. Mobility as liberty, mobility as progress. Everyday language reveals some of the meanings that accompany the idea of movement. We are always trying to get somewhere. No one wants to be stuck or bogged down.

   As he points out in his earlier book *On the Move*, this articulation reaches back to what is arguably the founding expression of a distinctly modern political imaginary, Thomas Hobbes's *Leviathan*, in which Hobbes, influenced by Galileo, presents the cosmos, including the world and its beings, as a system of objects in motion that rest only when forcibly stopped by external impediment (Cresswell 2006, 14–16). Motion, a natural state equated with freedom, is good; involuntary rest, ultimately equated with death, is bad. Hobbes's (1968) insight was that completely unregulated motion amongst human beings in social situations would lead to a proliferation of violent collisions. In the state of nature, unrestricted freedom of movement degenerates into its opposite: paralysis and death, and so Hobbes described the wager of society in terms of an artificial social contract, in which individuals exchange complete freedom and mobility for partial freedom and mobility, secured by a sovereign authorized to protect individual bodies in motion from other bodies in motion. Since the time of Hobbes, in both theory and practice, we have seen wide variation in the practical balance between individual freedom or mobility and the extent of sovereign authority, ranging from authoritarian situations in which the margin of individual freedom and mobility is thin, to liberal democratic situations in which the scope of individual mobility, rights and freedom is thought to be relatively broad by comparison.

The point is this: the moral valorization of mobility did not originate with the iPhone. The equation of freedom with mobility has been the central precept of the modern political imaginary from the outset. Accordingly, questioning the moral valorization of the mobile Internet in the contemporary context necessarily entails questioning the moral valorization of mobility itself in this imaginary, and while Hobbes (and many others after) might have been prepared to accept the equation of mobility and freedom as an objective, universal, scientifically established "fact," we know that that this proposition, and the norms that have been derived from it, were and are—like all knowledge propositions and norms—culturally produced and sustained. Indeed, it is the cultural and historical specificity of the ontological equation of the human with freedom, and of freedom with mobility (equations which, by the way, articulate very nicely with certain ideas about technology and market economies) that invites us to interrogate mobility as a norm that is contingent rather than necessary.

Among the things upon which the critical salience of the norm of mobility is contingent is the differential manner in which particular subjects or classes of subjects are afforded or denied it. Consider: the ongoing reality of denial of entry to asylum seekers at national borders; the threat of bodily violence that prevents women from moving safely through urban spaces; the spatial confinement of troublesome, typically racialized, people and populations by state authorities; the architectural denial of access to public spaces experienced daily by people with disabilities; the importance of mobile access to communication networks in uprisings against authoritarian regimes; and the isolation of senior citizens unable to get groceries or fill prescriptions because they cannot risk an icy sidewalk. Confronted with all this, categorical denial of the political urgency of mobility and its technologies is difficult to sustain. As Luc Boltanski and Eve Chiapello observe in their book *The New Spirit of Capitalism* (2005, 361), mobility is a crucial nexus of exploitation in highly networked economies: "In fact, in a connexionist world, mobility—the ability to move around autonomously not only in geographical space, but also between people, or in mental space, between ideas—is an essential quality of great men, such that the little people are characterized primarily by their fixity (their inflexibility)."

If mobility equals greatness and immobility poverty, then the prescription would seem obvious: get moving. However, the redistribution of certain forms of technologically enabled mobility and "flexibility" such that the little people might "enjoy" more of these seems to suit the interests of the great men just fine. How else to explain the consistency with which the merchants of transnational communicative capitalism express their claims upon our attention, bodies, money, time, creativity and imagination in terms of the imperative of incessant movement? Mobility—of information, communication and access to them; and of working people and consumers—and its contemporary technologies, are both culturally fetishized and essential structural conditions of contemporary economic and political power. Ours is a situation in which the experience of at least certain forms of mobility is relatively (though perhaps not perfectly equally) well-distributed, and in many cases even compulsory. This is the situation "enjoyed" by most of the working- and middle-class citizens of Euro-American capitalist liberal-democracies. With important limitations, exceptions and gradations indexed to age, gender, ethnicity and ability, these are people for whom both the experience and priority of mobility, especially as mediated by emerging information and communication technologies, is more or less *normal*. When Boltanski and Chiapello (2005, 361) write: "Great men do not stand still. Little people remain rooted to the spot," they express perfectly the "mobilist" ideology of networked capitalism. They also seem to appreciate the sociological ambiguity of this characterization. For it is far from clear that in responding to the ideological imperative to keep moving, the little people accomplish much beyond delivering themselves more effectively into their own exploitation by great men. As Boltanski and Chiapello (2005, 468) observe, when it comes to "loosening the grip of capitalism as an oppressive instance . . . One critical orientation, which is seemingly paradoxical given that mobility and liberation have hitherto been closely associated, is to be sought in challenging mobility as a prerequisite and incontestable value." Whereas a politics of distribution might recommend extending the presumed benefits of mobility and its technologies universally, a politics of disruption might instead require rejection of the very premise upon which this apparently egalitarian distributive inclination is based.

This proposition relies on a specification that associates politics with those activities by which an established horizon of consensus is disrupted. Such a specification is given by the French philosopher Jacques Rancière (2010a, vii), who characterizes consensus as the sense that "what is, is all there is." We live, Rancière says, in more or less consensual times. He (2010a, x) contrasts consensus with another way of being in the world, a way of being that "lays claim to one present against another and affirms that the visible, thinkable and possible can be described in many ways. This other way has a name. It is called politics." Rancière (2010a, 2) goes on to write that "Politics is the way of concerning oneself with human affairs based on the mad presupposition that anyone is as intelligent as anyone else and that at least one more thing can always be done other than what is being done." Becoming political means refusing to take the present state of things as given. It means disrupting the consensus that says that what is, is *all* there is, and that nothing can be done other than what *is* being done. Becoming political, as Rancière (2010a, 3) puts it, means claiming "the right to attend to the future." Such politics entail judgment and action that alter the parameters of the possible and the impossible in any given situation. It is the sort of politics that can be distinguished from what Rancière (1999, 28–30) elsewhere calls "police," referring to those agencies, practices and institutions—including the institutions of liberal democratic government— whose function it is to contain the disruptive possibility of politics, even as they give the impression of it.

In what relation to this sort of politics do mobility and its technologies stand? As discussed above, mobility is foundational to modern western conceptions of freedom. Another conspicuous aspect of Western modernity is the promise that freedom-as-mobility can be delivered by technology. Modern western culture has thus reserved a special place in its imaginary for transportation technologies—trains and railways, automobiles and highways, airplanes and skyways, rockets and space travel—that were supposed to deliver on the promise of freedom as technologically enabled movement through space. But no mere transportation technology could ever truly fulfill the ultimate dream of mobility: the dream of being in two places at one time. It was only when electricity supposedly made it possible to liberate communication from its reliance on transportation that progress toward this dream began in earnest, via a succession of communication technologies and accompanying rhetorics that have culminated in contemporary digital networks and loose talk about the annihilation of distance, time-space compression, the empire of speed and interactivity in real time.

As Jonathan Sterne (2006) has persuasively argued, communication and transportation are perhaps not so easily, or so advisedly, separated as either James Carey's canonical account of the telegraph or residual preoccupations with the symbolic over the material dimensions of communication would have us believe. It is thus commendable that the "mobilities paradigm," as it has been formulated by John Urry (2007, 147), includes the corporeal travel of bodies and the physical movement of objects (i.e., transportation) as well as the imaginative, virtual and communicative movement of symbols and representations. Scholars of mobility know very well that mobility entails both transportation and communication (perhaps the missing term here is *mediation*, of which both transportation and communication are forms). In the popular (and certainly the commercial) imaginary, freedom-as-mobility specifically implies a dream of spatial transcendence (the dream of being in two, or more, places at once) that emerging information and communication technologies are promoted as uniquely configured to deliver. It is in this context that something like the moral valorization of the mobile Internet begins to make sense. If freedom is identical to transcending the limitations enforced upon the movement of our bodies through space, then maybe with mobile technologies we really are there.

Unless, that is, mobility and freedom are actually about time, not space. Significantly, at the very moment Boltansi and Chiapello (2005, 468) begin to consider the possibility of resistance to, and liberation from, the "new spirit of capitalism" in which mobility and its technologies play a central role, their attention shifts from the spatial to the temporal register. "The first problem, an absolutely concrete one," they write, "concerns the use of time." They elaborate: "Maybe a step in the direction of liberation today involves the possibility of slowing down the pace of connections, without thereby fearing that one no longer exists for others or sinking into oblivions and, ultimately, exclusion." It is at this moment that the possibility of a disruptive politics of immobility suddenly appears on the horizon. In a lecture at the 2009 conference on "The Idea of Communism" held at the Birckbeck Institute in London, Rancière (2010b, 168) said something striking about the temporal dimension of a specifically egalitarian form of freedom. "The emancipation of the workers," he said, "means the affirmation that work can wait."

Work can wait. Two things are happening here. The first is that freedom is identified as a question of the division of labor, and is located specifically in the experience of workers and their work. The second is that the question of freedom is here registered as a question not of transcendence or control over space, but control over time, the time of work, and just as mobile information and communication technologies bear on the promise of the former, so too do they bear on the possibility of the latter. We know that struggles over the time of work—the duration and structure of the workday and week; the age at which people begin and end their lifetime of work; the pace, speed and rhythm of production; the value of a waged hour—have perennially been a focal point for workers' struggles to recover something of freedom in the context of capitalist relations of production. We also have a wealth of outstanding critical scholarship concerning the manner in which digital technologies have been intimately involved in the proliferation of restructured "flexible" work arrangements that can hardly be said to have finally delivered emancipation to people who work. The question is whether technologies of mobile access to information and communication networks can reasonably be said to increase the chances that working people might be emancipated from the temporal demands of work. This is an empirical question whose definitive answer would have to reckon with the potentially great diversity of individual experiences of work in the mobile, networked economy. However, it seems that whether we are talking about Terranova's (2004) free laborers, Lazzarato's (1996) immaterial laborers, or professional, creative, administrative, clerical or service workers of any type (Head 2003), it would be difficult to believe that technologies of the mobile Internet have *increased* their ability to contain the time they spend working. While mobility may make it possible for people to choose to work *wherever* (and therefore *whenever*) they want to, this is quite the opposite of an emancipated situation in which the work can *wait*, and it serves only to illustrate the political bankruptcy of the concept of choice. Every day we are surrounded by people who "choose" to work incessantly, not because mobile technologies mean that they *can*, but rather because the mere availability of these technologies suffices to make them accept that they must. The work cannot wait because mobile technology means it does not have to. The email, the unsorted post, need not pile up. Precisely because it delivers on the spatial dream of being in two places at once, the mobile Internet undermines the temporal dream of a day when the work can wait.

What might we expect from politics under technological conditions where work cannot wait? About a year ago I was invited to give talk at a university in Canada and had the occasion to sit down with a friend and colleague whose work I greatly admire. He was (and is) a genuine left intellectual, having been in the streets of Paris in May 1968 and having studied at the feet of Herbert Marcuse. I knew he kept a flat in Paris, and so I asked him if he knew about the Tarnac Nine, whose case I had recently become fascinated with. They were are a group of well-educated, middle-class young people who had read radical philosophers and moved to the mountain village of Tarnac in the Correze region of central France, where they established a communal farm, delivered food to the elderly and infirmed, reopened the general store as a cooperative, and established a local film society and lending library. In 2008, nine of these young people, now known as the Tarnac Nine, were arrested on terrorism charges, accused of sabotaging power lines in an act that threw high-speed train service around Paris into chaos for several hours. When I asked my colleague what he thought about this he was immediately visited by the specter of inconvenience: "That's terrible," he said, "disrupting the transit system just makes it hard for people *to get to work*." After all, work cannot wait. Or maybe it can. This is what the normative expectation of mobility, the experience of mobility as *normal*, offers to the possibility of politics conceived under the sign of a disruption that alters the distribution of possibility and impossibility: a target.

This is the insight and lesson of the Tarnac Nine. The manifesto *The Coming Insurrection*, written by the Invisible Committee and widely attributed to the group, includes a scorching critique of the contemporary capitalist state and culture in France, and a call to political action that turns precisely on the question of mobility and its relationship to work. In the contemporary milieu, they observe, mobility is a condition not so much of work itself as of employability, of being constantly ready and available to work. Describing the networked system of flexible production as "a gigantic apparatus for psychic and physical *mobilization*," they observe: "Mobility brings about a fusion of the two contradictory poles of work: here we participate in our own exploitation, and all participation is exploited"

(Invisible Committee 2009, 50–51). Under these conditions, nothing could be more abnormal or threatening than idleness or unavailability for occupation. As they (2009, 48) write: "The menace of a general demobilization is the specter that haunts the present system of production."

In circumstances such as ours, where mobility is considered normal and normative, demobilization constitutes the sort of disruption that can open the field of political possibility. In this instance, not only the symbolic "ethos of mobility," but also the *material infrastructure* of mobility, becomes a priority target against which to enact this disruption. Invoking the potential of "a single incident with a high-voltage wire," they write that: "In order for something to rise up in the midst of the metropolis and open up other possibilities, the first act must be to interrupt its *perpetuum mobile*" (Invisible Committee 2009, 61). When Michele Alliot-Marie, the French Minister of the Interior who orchestrated the spectacular arrest of the Tarnac Nine, held up the fact that "they never use mobile telephones" as evidence of their "pre-terrorist" tendencies, she meant to say that anyone who wants to evade surveillance must be guilty of something, but she dared not speak the truth of the group's more radical antagonism toward mobility itself (quoted in Toscano 2009, n.p.). This was an antagonism that, by its enactment, represented the sort of disruption that could make real the possibility that work can wait, and so loosen the grip of mobility's normativity on the time of politics. Their purpose was not only to evade capture, but to sabotage mobility itself, as a condition of the possibility of politics:

> The technical infrastructure of the metropolis is vulnerable. Its flows amount to more than the transportation of people and commodities. Information and energy circulate via wire networks, fibers and channels, and these can be attacked. Nowadays sabotaging the social machine with any real effect involves re-appropriating and reinventing the ways of interrupting its networks. How can a TGV line or an electrical network be rendered useless? How does one find the weak points in computer networks, or scramble radio waves and fill screens with white noise?
>
> (Invisible Committee 2009, 111–112)

In their own more succinct words in relation to normative mobility, theirs was a politics of "fucking it all up" (Invisible Committee 2009, 112).

To those for whom the prospect of politics is properly contained within the normative framework of liberal publicity (intersubjective public dialogue supported by freedom of information and freedom of expression), such a position cannot sound anything but extreme. In situations where publicity is systematically denied by authority, access to systems of information, communication and mobility are crucial to the possibility of political judgment and action. In these cases, blank screens and blocked transit systems are tools of a regime desperately trying to hold on to power. On the other hand, for those whose situation is saturated by liberal publicity but conspicuously devoid of politics, and who therefore think the possibility of politics relies upon exceeding the affordances of publicity, the normativity of mobility presents an interesting opportunity. In situations where transportation and communication collapse into systems of mobility that bind us to work that cannot wait, loosening the grip of power might require the sort of letting go that is characteristic of paralysis. Paralysis, the loss of the ability to move, the inability of a system to function properly, derives from the Greek *paralusis*, whose root is the verb *luein*, which means to loosen, untie or release. In paralysis, unable to move, we might be released from that work which cannot wait (and the power it represents and materializes) to which we are otherwise bound by the normative imaginary and materiality of mobility. In losing mobility we might at last come undone, and find ourselves on the undecideable, unpredictable, unfamiliar terrain of politics.

In a culture that identifies movement with freedom, and immobility with powerlessness and death, it is hard to make the case for paralysis as opening onto the possibility of politics and the politics of possibility. We are embarrassed when Willie Corduff and his neighbors—the "little people" rooted to the spot on which they stand—say they are "going nowhere" because, after all, the gas does need to flow just as surely as people do need to get to work. For us, "This could go on forever" is an expression of frustration, not agency: it is what we say when we are stuck in a line that does not seem to be moving. Perhaps we misreckon the radical potential of paralysis in circumstances where power is enacted through symbolic and material infrastructures of mobility that together ensure that work cannot wait. Referring to some pre-Internet systems of mobility—garbage collection, the postal system, electric street lighting—in his elegant essay "Sir, Writing by Candlelight," E. P. Thomson (1980) expresses this potential when he writes: "It is only when the dustbins linger in the street, the unsorted post piles up—it is only when the power workers throw across the switches and look out into a darkness of their own making—that the servants know suddenly the great unspoken fact about our society: their own daily power."

SOURCES

Boltanski, L., and E. Chiapello. *The New Spirit of Capitalism*. London: Verso, 2005.
Cresswell, T. *On the Move: Mobility in the Western World*. London: Routledge, 2006.
Cresswell, T. "Towards a Politics of Mobility." *Environment and Planning D: Society 16 and Space*. no. 1 (2010): 17–31.
Domhnaill, R. "Scanna'in Inbhear/Underground Films/ Riverside 8 Television." *The Pipe*. 2010. Accessed February 1, 2014. www.thepipethefilm.com
Head, S. *The New Ruthless Economy: Work & Power in the Digital Age*. Oxford: Oxford University Press, 2003.
Hobbes, T. *Leviathan*, edited by C. B. Macpherson. New York: Penguin, 1968.
Invisible Committee. *The Coming Insurrection*. Los Angeles: Semiotext(e), 2009.

Lazzarato, M. *Immaterial Labour. Radical Thought in Italy: A Potential Politics,* edited by Hardt and Virno. Minneapolis: University of Minnesota Press, 1996.

Rancière, J. *Disagreement: Politics and Philosophy,* translated by Julie Rose. Minneapolis: University of Minnesota Press, 1999.

Rancière, J. *Chronicles of Consensual Times.* London: Continuum, 2010a.

Rancière, J. *Communists without Communism. The Idea of Communism,* edited by Douzinas and Žižek. London: Verso, 2010b.

Siggins, L. *Once Upon a Time in the West: The Corrib Gas Controversy.* Dublin: Transworld Ireland, 2010.

Sterne, J. "Transportation and Communication, Together as You've Always Wanted Them." In *J. Packer and C. Robertson. Thinking with James Carey: Essays on Communication, Transportation, History,* edited by Packer and Robertson, 117–136. New York: Peter Lang, 2006.

Terranova, T. *Network Culture: Politics for the Information Age.* London: Pluto Press, 2004.

Thomson, E. P. "Sir, Writing by Candlelight." In *Writing by Candlelight.* London: Merlin Press, 1980.

Toscano, A. "Criminalizing Dissent," Guardian.co.uk. January 28, 2009. Accessed February 1, 2014. www.theguardian.com/commentisfree/libertycentral/2009/jan/28/human-rights-tarnac-nine

Urry, J. *Mobilities.* London: Polity, 2007.

2 Openness and Enclosure in Mobile Internet Architecture

Alison Powell

Most histories of the Internet stress the alignment between an open architectural form and the 'open' or democratic values of some of the earliest Internet pioneers. This continues today as we celebrate connections between Internet technology, freedom and democracy. The reality, however, is that the architecture and experience of the Internet are changing. There are many ways in which these changes happen: technical standards change; nations become more interested in regulating or controlling the flows of Internet traffic; and legal standards shift the possibility of interconnection between networks is governed. One particular change that often doesn't receive much attention is the shift from the 'wired' Internet to the 'mobile' Internet, which occurs materially through our increased use of mobile devices, as well as through a shared imaginary that increasingly frames the Internet as being accessible anywhere and everywhere via mobile devices. This means that the way we think about our Internet experience—as seamless, everywhere and always on—relies upon the way that our Internet experience is being built, which is quite different in the 'mobile' space than it was in the 'wired' one. This shift has implications for user experience and practice, but it also has political-economic implications that are linked with the different history and technical architectures that are historically part of the mobile, rather than Internet, industry. The shift also calls into question the extent to which the openness that we tend to associate with the Internet results from a very particular historical confluence of factors.

This chapter explores how this shift allows us to analyze the way that openness—as architecture and practice—has been imagined in the past, and examine the factors that might enable or constrain it in future. As such, it explores the amazing persistence of imaginaries of openness even as they overlap and conflict with materialities of enclosure. These overlaps and conflicts provide a context for some recent history of the Internet, as well as chilling premonitions of its future. As the material platforms and architecture that support the Internet experience move to mobile devices, various types of enclosure can occur that are not only technical but which depend upon social norms and practices. These include the way that mobiles employ proprietary protocols and standards for interconnection, as well as the differences between the 'open-source ecology' of peer production of Internet software and the 'app ecology' of mobile application development. Their interrelationship suggests that we may need to change our expectations of openness or participation in building the Internet.

This chapter reviews the origins of the 'open' Internet, placing particular attention on the way that particular values were imagined through the design of Internet networks. These articulations between design and politics set expectations about the importance of features such as 'openness.' Openness is associated with increased opportunities for participation in altering the Internet itself. This participation is part of what Zittrain (2008) perceives as the 'generative' potential of the Internet. Free and open-source software development is the exemplary case of this participation, since it uses the Internet to organize a group of people whose actions then modify the material-technical form of the Internet. These 'recursive publics' have significantly contributed to the imaginary of the 'open' Internet as being especially participatory. The second section contrasts this historical 'wired' Internet with the increasingly ubiquitous 'wireless' Internet, examining the overlap and tension between how social aspects of the technology are imagined, and how it is designed, regulated and otherwise materially constituted. The section begins by examining the differences in the materialities of mobile Internet access devices and the 'wired' Internet, and continues by examining how the participatory cultures emerging in relation to the mobile Internet are different than the 'recursive publics' of the wired Internet. Opportunities to 'open' mobiles exist, but in many cases they are limited by material, legal and organizational structures.

BACKGROUND: IDEALS OF INTERNET OPENNESS

The history of the Internet and especially of the idea of the 'open' Internet illustrates the way that particular social values can become articulated—connected—with technological design features (see Slack 1997). It also demonstrates how particular technological affordances can lay the groundwork for the emergence of particular social and cultural forms. Along with the technical choices made about specific technologies, normative and cultural aspects also combine to shape the forms of communication that emerge. We can think of the Internet as a form of communication that is formed of technical structures as well as of organizational frameworks, notions of governance, and images of the future. Many histories of the Internet have acknowledged this socio-technical coproduction, examining how existing collaborative processes of university research scientists contributed to the network structure of the early Internet (Abbate 1999) as well as how elements of a 'hacker culture' that emerged in the MIT Artificial Intelligence lab in the 1960s influenced the cultures of production that came to characterize the Internet.[1]

These specific reflections contribute to a broader analysis of the interplay between social and technical imaginaries of the Internet, or what we in this book consider as the 'materialities and imaginaries.' This interplay is particularly interesting to scholars working at the interface of media and communication technologies and the tradition of science and technology studies (STS). Broadly speaking, this intellectual space is concerned with the *coproduction* of ideas about things, the social practices established around things and the things themselves. Patrice Flichy (2007) argues that the social imaginaries of new technology are as strongly influenced by the way that people think about and construct them as they are by the technology itself. Technologies therefore operate as sites of knowledge transfer or exchange (Bowker and Star 1999) and as elements of controversies that mobilize opposing social or cultural perspectives (Callon 1981). In one of the defining works on coproduction, Jasanoff (2004, 2) defines it as being "shorthand for the proposition that the ways in which we know and represent the world (both nature and society) are inseparable from the ways in which we choose to live in it." From this perspective, we can see how the material infrastructure of technologies like the Internet is produced along with the symbolic infrastructure of the social. In the past, this symbolic infrastructure was focused on individual liberty. More recently, it has become associated with participation: through the participatory 'Web 2.0' and also through various open-source movements.

The coproduction of the technical, cultural and normative aspects of the 'open' Internet are well illustrated by the way that specific ideas about the social value of openness have been connected with specific technical choices, and secured through norms such as standards and protocols. These three elements are key to how the 'open' Internet has developed.

Technology and Culture: The Internet's Libertarian History

One key feature of previous and current imaginaries of the Internet has been the significance of the political and cultural values articulated with the design and function of a 'network of networks,' which can conflict and overlap with how such a network is conceived of in its ideal form. These values have often been identified as 'freedom' and 'openness,' although these are not exactly identical. As an illustration, consider the names of some of the more prominent nongovernmental organizations (NGOs) active in lobbying for digital rights: these include the Electronic Freedom Foundation, the Open Network Initiative and the Open Technology Initiative of the New America Foundation. These evocations of the Internet's liberty and freedom have long histories. As computer historians and sociologists, including historian Fred Turner, have noted, the cultural values of Californian Internet geeks have tended toward the individualist and libertarian (Turner 2006). In particular, 'openness' has been a key point of contact between peer-to-peer cultural practices and neoliberal ideologies, although it is often problematically conflated with individual freedom. In her 2013 work exploring how software production coproduces politics and code, Coleman analyses how the individual freedom that free software coders sought and developed through their individual practices in some ways operated as 'theatre of proof' for a broader openness of 'artistic, academic, journalistic, and economic production' (2013, 185). This unstable and yet productive relationship between the political ambiguity of free software and the replicability of its norms and values across a number of contexts has made it very influential. The idea that a reprogrammable Internet is an 'open' Internet has wound its way through discourse about the technology from the early age of the Web, maintaining expectations that the Internet provides space for its own construction.

The fascination with the 'open' Web also has historical roots in the conflation of libertarian values and technical architecture. Take for example, John Perry Barlow's "A Declaration of Independence of Cyberspace," where he writes,

> Governments of the Industrial World, you weary giants of flesh and steel, I come from Cyberspace, the new home of Mind. On behalf of the future, I ask you of the past to leave us alone. You are not welcome among us. You have no sovereignty where we gather. We have no elected government, nor are we likely to have one, so I address you with no greater authority than that with which liberty itself always speaks. I declare the global social space we are building to be naturally independent of the tyrannies you seek to impose on us. You have no moral right to rule us nor do you possess any methods of enforcement we have true reason to fear.
>
> (Barlow 1996, n.p.)

This libertarian discourse focused on the specificity of the online environment relative to the jurisdictions over which states had governing responsibility. This perspective was typical of mid-1990s perspectives assigning the Internet special status by virtue of the architecture and function of its technology. Barlow continues,

> Cyberspace consists of transactions, relationships, and thought itself, arrayed like a standing wave in the web of our communications. Ours is a world that is both everywhere and nowhere, but it is not where bodies live.
>
> (Barlow 1996, n.p.)

These claims that the Internet has particular features that make it impractical or ideologically problematic to regulate (like a network structure and a global reach) have been remarkably persistent. Hofmann (2009, 2) argues that these claims are part of a "utopian vision of autonomy and creativity" that contrasted traditional, switched telecommunications with a distributed, autonomously connected network of networks: an *Internet*. Where the original switched network embodied a "bureaucratic model which emphasizes collective security, stability and regularity" (Hofmann 2009, 5–6), the utopian imaginary of the Internet suggested an alternative where the intelligence belonged to the users of the network, who could use the noncentralized network to communicate without limits.

The political expectations of the development of an Internet as opposed to centralized and switched network are expressed in a set of essays by David Isenberg in which he lauds the potential of the "stupid network" in which interconnection is valued over the "intelligence" of pre-Internet telephone networks. Isenberg thought of the stupid network as a philosophy, writing that "A new network "philosophy and architecture" is replacing the vision of an Intelligent Network. The vision is one in which the public communications network would be engineered for "always on" use, not intermittence and scarcity. It would be engineered for intelligence at the end-user's device, not in the network, and the network would be engineered simply to "Deliver the Bits, Stupid" (Isenberg 1997).

In a 2002 update of the "stupid network" idea, Isenberg, along with David Weinberger, argued that managed "best networks" would be less beneficial than open networks. This technical recommendation (itself based on a business case assumption that open networks would include ever-increasing amounts of bandwidth) includes an explicit political proposition:

> In fact, the best network embodies explicit political ideals—it would be disingenuous to pretend it didn't. The best technological network is also the most open political network. The best network is not only simple, low-cost, robust and innovation-friendly, it is also best at promoting a free, democratic, pluralistic, participatory society; a society in which people with new business ideas are free to fail and free to succeed in the marketplace.
>
> (Isenberg and Weinberger 2002, n.p.)

Political openness and technical openness are explicitly aligned here. In contemporary imaginaries of the Internet, we find the same sentiments, expressed through the names of the NGOs mentioned above. We can also find rhetoric explicitly linking open networks and democratic political values in the language of lobbyists advocating in favor of net neutrality regulations (Powell and Cooper 2011) and in the political discourse of political leaders including Hillary Clinton and David Cameron (Powell 2011). The salience of this imaginary of openness, however, is also solidified through social norms including a tradition of peer governance of the Internet that, in part, has occurred through the development of technical standards and protocols. Like the free software programmers that Coleman discusses, the early hackers, geeks and technocrats who established the protocols over which the Internet operated (including Isenberg, Barlow and others) also established mechanisms for the debate and approval of those protocols, which allowed the network to be designed in ways that echoed the political and social interests and values of this group.

Open Protocols and Peer Governance: The 'Wired' Internet's Norms

The early Internet was designed as an open and interoperable network, with the 'intelligence' at its edges, as Isenberg pointed out. This openness was secured by the core standards and protocols governing the function of the network. The Internet Protocol, along with the Transport Control Protocol, together prescribe the interconnection of the Internet.[2] These protocols were originally developed by engineers David Clark and Vint Cerf but each new version of these—and any other technical aspects of the Internet protocol— are presented to the worldwide community of Internet designers through documents called Requests for Comments.[3] Braman (2011) notes that these documents, in addition to defining technical standards through review by the entire community of Internet designers, also provide sites where some of the most important political issues of the early Internet were discussed.

The 'wired' Internet operates on standards and protocols which, in many cases, have been agreed through a process of community review, and codification has been a key part of the technical design of the 'wired' Internet. The review by designers has also been a key part of its governance. Technical decision-making regarding the standards and protocols is the responsibility of standards-setting bodies, notably the Internet Engineering Task Force (IETF).[4] This task force is open to anyone who wishes to attend its meetings, and is in theory a nonhierarchical organization that bases its decision-making on "rough consensus and running code." Eschewing voting, decisions are made by consensus, sometimes even by asking participants to hum (and then considering the volume of humming in the room as an indication of 'rough' consensus). As for the running code, a key aspect of the original governance of the Internet was that many decisions about how data moved and communications were put in place were delegated to the protocols that had been agreed by the community. If the code or protocol worked (technically) that was a convincing argument for continuing to use it to build the Internet. In combination, these technical and normative aspects combined to create a de facto system of peer governance of the Internet, in which technically skilled Internet designers made decisions about how the Internet ran, that they hoped would be in tune with their values.

Peer governance expanded along with the Internet, although not always without difficulty. These transnational advocacy networks, as Mueller (2010) points out, deliberate across borders, and have established international institutions such as the Internet Governance Forum (IGF), a United Nations-supported multi-stakeholder meeting which was itself established after several World Summits on the Information Society (WSIS).[5] These institutions deliberately engage multiple stakeholders beyond the technical experts whose 'rough consensus and running code' governed the Internet to begin with. The WSIS and the IGF, in particular, have attempted to expand governance decision-making to include not only states and corporate entities but civil society as well. As Held and McGrew (2003) note, this process was originally meant to shift the form and the process of governance from one based on confined, state-level decision-making toward negotiation across a variety of levels. "Multistakeholder governance means that representatives of public interest advocacy groups, business association or other interested groups can participate in intergovernmental decisions alongside governments" (Mueller 2010, 7–8). Raboy notes that these decision-making fora "form the basis of a new model of representation and legitimation of non-governmental input to global affairs," and as a result, "the rules and parameters of global governance have shifted" (Raboy 2004, 349).

Taken together, the new transnational and nonhierarchical institutions that govern the Internet introduce opportunities to make the Internet a technology managed by its peers. Thus, the standards and protocols that technologically defined the Internet have also, over the past decade, helped to reiterate and develop shared cultural norms.

The 'Wired' Internet's Alternatives—and Their Articulation to Dominant Modes of Engagement

Besides the design of standards and protocols, and the human systems of governance that order them, much of the software that supports the function of the Internet is also constructed as 'open'—because it is developed and licensed as free or open-source software (FLOSS). 'Share-alike' licenses, including the GNU Public License (Stallman 1999), stipulate that software products based on free software code must be released to a common repository for reuse. The Free Software Foundation's goals were to define, using the framework of copyright law, a mechanism through which software code produced from free software source code would always be available for community reuse. In a way this alternative approach reiterates the kind of peer governance developed by the IETF, which claimed 'running code' to fulfill the same governing purpose as consensus on the definitions of new standards or protocols. In time, the governance of free software also gave rise to the expansion of open-source software practices, although not all open-source software licenses have stipulated that code must be returned to commons. To an extent, FLOSS alters the hierarchical structure of the traditional firm (see Stark 2006) by allowing individuals to self-organize and, in the case of free and open-source software, to create collective repositories of common knowledge. Increasingly larger corporations have begun to integrate open-source production into their business models, further blurring the lines between the commons and the proprietary, and creating more complex integration between horizontal and hierarchical organization (Benkler 2006; Coleman 2013). This facilitation of participation and maintenance of a knowledge commons again creates greater opportunities for shared governance, as Bauwens (2009) points out. The movement of FLOSS practices also illustrates how imaginaries—and indeed ideologies of 'openness'—are transported through certain communities of practice to become relevant in locations and contexts in which they had not been before. In some of these new contexts, the original values connected with 'openness' shift. In the case of open source, the recasting of free software as "open source" and the creation of licenses that no longer required the return of software code to a commons created a linkage between the commons and the market, reconfiguring the software industry to align with Benkler's "commons-based peer production." This 'openness' shifts from an openness imagined through libertarian ideals of freedom, and toward the kind of openness associated with liberal, free markets. This expansion and mutation of the value of Internet 'openness' accompanied the commercialization of the Internet.

The 'wired' Internet has carried forward design and governance features that link with the historical imaginaries of its openness. Now, various changes are impacting both the design and the governance of Internet resources, leading to a tension between the materialities and the imaginaries of the Internet. Some of the points of tension include pressure from national governments to censor or limit certain Internet services, and arguments from Internet service providers and other corporate actors to allow for more prioritization of the data moving across the networks, undermining the original design principle of "network neutrality" (Wu 2003). Yet very often, one of the most significant challenges to the openness of the Internet is not because of any individual regulatory decision but rather as a result of a large-scale shift in the modes of Internet access.

The Move to Mobile: Shifts in Cultural Norms

The increasing use of mobile devices for Internet access is a significant factor in the experience and framing of Internet openness. According to ComScore's 2013 data, 125 million people in the USA use smartphones, and these devices account for 50 percent of mobile phones in that market (Lipsman 2013). The Pew Internet and American Life project estimates that 63 percent of adult mobile phone users go online with their phones, and 21 percent of adult mobile users do most of their Web browsing from their phones (Duggan and Smith 2013). Besides access to Web pages, mobile data services include email and applications, maps and Internet telephony such as Skype. In this shifting context we can see that 'Internet' use on mobiles is somewhat fluid: applications require data connections but not necessarily a connection to the global Internet, while email, even though it is transferred across the public Internet from its source, is delivered to mobile devices through proprietary protocols.

In the context of these shifts, it is significant that some materialities, including policies and protocols for mobile devices, are far less 'open' than those that developed around the 'wired Internet.' For example, in the United States, the Federal Communications Commission (FCC) recently ruled that telecommunications companies needed to preserve network neutrality—except in mobile markets. This establishes a systematic shift in materialities away from openness and back toward some of the modes of control that were more common in the kinds of 'smart' communication networks that Isenberg pilloried in his original texts. Less attention is paid to these shifts, in part because of the way that the Internet has increasingly become an essential infrastructure—and hence, invisible. As Star and Bowker (2002) point out, it is the point at which infrastructures cease to become visible (that is, the point at which they become banal) where they have the most power. This represents the paradox of communication ubiquity: as Internet connectivity becomes increasingly expected, it becomes less clear how that connection actually occurs. As we will see, this creates the opportunity for a significant departure from the forms of governance developed by the 'fixed' Internet's designers: the original Internet protocols were based upon open standards, so they worked the same way on a range of devices.

MOBILE ENCLOSURE

Other aspects of mobile materialities also play a role. Mobile devices have been developed in a different institutional mode than the original 'open' Internet, meaning that the architectural aspects as well as the standards and protocols are more constrained. The next section reviews how these constraints operate, before moving on to discuss the extent to which alternative cultures of production attempt to reinvigorate imaginaries of openness in the face of these constraints, through activities like 'open-sourcing' mobile phone production.

Technical Enclosure: Architectural Limits on Openness

The historical organization of the mobile industry has influenced the architecture used by mobile Internet access devices. Historic competition between device manufacturers inspired the development of proprietary rather than open standards and protocols. Telecommunications providers, as compared to Internet service providers, have historically provided (and often owned) the access devices that their customers used. For example, the BlackBerry mobile device was one of the first to develop mobile e-mail; it uses a proprietary protocol and its own server systems to deliver email to mobile devices as soon as it arrives. In contrast, the IETF's Internet Message Access Protocol (IMAP) and Simple Mail Transfer Protocol (SMTP) protocols, which are open, are used for the same purpose on the 'wired' Internet across a range of devices. The allocation of radio spectrum to mobile operators by auction also contributes to the increasingly enclosed mobile architecture. Mobile companies bid on exclusive licenses to portions of the radio spectrum, meaning that mobile data, unlike Internet data, moves across a number of separate networks instead of one 'network of networks.' This seems like a potentially minor point, but the necessity for interconnecting many networks was one motivating reason for the development of interoperable and platform-agnostic Internet standards and protocols. In contrast, the single networks of the mobile operators represent a return to control of the entire process of data transmission—from sender to receiver. This is quite distinct from a 'best effort" network like the Internet, where messages are separated into packets that are routed in various ways to their destination, according to well-defined protocols (i.e., TCP/IP). The principle of 'net neutrality' that underlies this form of data delivery has been protected on the 'wired' Internet by the FCC's recent ruling, but not for mobile networks, who are permitted to block content or engage in any other kinds of control of content that they wish. These forms of technical enclosure, themselves linked to emerging policy frames, are the first example of mobile enclosure.

Consumer Choice

A second aspect concerns the way that users of mobiles experience limits to the kinds of devices they can choose, and the kinds of things they can do with them. Baer et al. (2011) compare several features of mobile device openness across the economies of Brazil, China, India, France, the United Kingdom and the United States, including consumer choice, usage restrictions and the scope for innovation coming from device makers, application developers and mobile operators. They conclude that mobile ownership is highly consolidated, with each major market having only three or four major players. Devices are often tethered to specific operators, and consumer choice is limited by long-term contracts and restrictions on device portability. They write, "Restrictions on device portability place limits on what users can do with their phones—preventing their use on competing carriers' networks or on foreign networks when traveling, and denying access to certain apps" (Baer et al. 2011, 34).

Beyond these consumer issues, however, there are more significant concerns about the extent to which applications and the market for applications have influenced the openness of mobiles. One of the most significant changes to the experience of using mobile devices has been the development of applications. Introduced in 2008 with five hundred applications on offer, as of 2012 the Apple App Store hosted 450,000 applications, according to CEO Tim Cook (Indvik 2012). Before the introduction of these applications, manufacturers preinstalled applications like calendars onto mobile devices. The application market has allowed third-party developers (sometimes smaller companies or individuals) to produce applications for Apple's iOS operating system or Google's Android operating system. Comprising an effective duopoly in mobile applications, at least at the time of writing, these companies control the approval of applications that run on their platforms. Apple, in particular, has an extremely stringent approval process, refusing applications that it claims violate community norms. The company also restricts visitors to the applications stores located in their own countries. Instead of privacy and security issues being addressed through regulation, as they are within the historically regulated telecommunications sector, they are addressed through corporate policy and communicated to subscribers through often lengthy and detailed terms and conditions.

In November 2011, Apple stipulated that all developers wishing to build applications for the iOS will be required to add an extra layer of security to the wares they develop, which limit the extent to which these applications interact with the operating system (Goodin 2011). Apple and Google often stipulate that data produced by someone who uses the applications can be gathered, retained and used by the parent company—though not by the application developer. This means that app developers cannot easily benefit from the data developed through their applications. App developers are also bound by the terms and conditions set for them by the owners of the platforms, set out in software development kit licenses (SDKs)[6] that define how applications must be programmed in order to work properly. These SDKs can be highly restrictive. Ironically, Google's Android platform is based on an open-source software stack, yet its SDK clearly stipulates that the finished app is actually the property of either Google or Android. In contrast to the "recursive publics" (see Kelty 2005) of Internet users who built software to improve the function of the Internet, and who established some principles of peer governance through involvement in standards-setting, the app economy is a one-way process where the products created by app developers help to secure valuable data for the mobile phone companies, add to the value of the devices and cede developers' code to the device owners. For their part, developers are paid for their creations, and may gain reputation and status by being associated with a particularly well-made application. However, unlike the 'open' contribution structures of the wired Internet, which invited not only contributions to various open-source projects where code is maintained for use by others but also contributions to the development of open standards, the ecology of application development is far less open. In particular, it employs the logic of contribution without the logic of the commons.

Open Alternatives? The Case of Openmoko Phones

The move to a mobile Internet returns many forms of corporate control to the communications ecosystem, and adds new forms such as the management of the app economy through the solicitation of contribution without the facilitation of peer governance. Yet alongside these enclosures of common knowledge and architectural shifts away from openness, alternatives are also emerging. The final example I consider is of a project that aimed to completely open-source the development of a mobile phone, examining its aims and goals, and the eventual outcomes of its aims.

The Openmoko project was an attempt to create an entirely open mobile phone, including hardware, software and governance of the production process. This project was led by Sean Moss-Pultz, who was a product manager at a Taiwanese company that produced motherboards and other electronic hardware. In the press release that accompanied the launch of the project in 2006, Sean is quoted as saying, "For the first time, the mobile ecosystem will be as open as the PC, and mobile applications equally as diverse and more easily accessible . . . Ringtones are already a multi-billion dollar market. We think downloading mobile applications on an open platform will be even bigger" (Openmoko 2007). The project aimed to create a common platform for mobile application development, based on the Linux open-source operating system, as well as common storage models and libraries for applications. It solicited participation from open-source software developers, people engaged in building mobile applications, and hardware hackers as well.

In early 2007, Openmoko launched the Neo1973, the world's first open-source smartphone. With the launch of their first product, the project expanded the scope of their open-source aims to include a product development process that explicitly involved participation from a group of motivated amateurs—much like other open-source projects, but unusually, also including the company's development roadmap. In February 2007, the project's website included the following outline of the process:

> Our company is unconventional, we openly share our roadmap, and your participation, in terms of actual code, hardware features, suggestions, and usage-scenarios will shape product features of our future devices.
>
> Each product we build follows a three-phase development model that repeats annually:

Phase 0: Developers Preview

> We will give away free phones to selected members of the developer community. At this point, the full source code to the product will become publicly available. We are committed to cooperating with the community in the interest of making the official developer launch a smooth experience. Also, at this time, all the community websites will be updated with content from the new device.

Phase 1: Official Developer Launch

> We will sell the Neo device direct from openmoko.com. Sales and orders will be worldwide. We are specifically targeting open source community developers.

Phase 2: Mass Market Sale

Online sales will continue. We will also be available in a retail stores and selected carriers around the world. At this point, we hope your mom and dad will want to buy a Neo, too. The next generation Openmoko device will also be introduced at this time.

We want your involvement in Openmoko. Now is a great time for us to work together. You'll have our full support. We're dedicated to helping you "Free Your Phone." And we're always looking, listening, and hungry for new things. It is our goal to be totally market driven.

(Openmoko 2007, n.p.)

On the community's mailing list, Moss-Pultz repeated the same message, adding:

Our goal is freeing end-users and businesses alike from proprietary constraints. We're about encouraging people to modify and personalize their software to support their individual needs.

Building products as we do, we strive to enable people to connect and communicate in new and relevant ways, using their own languages and their own symbols. . . . Your participation, in terms of actual code, hardware features, suggestions, and usage-scenarios will shape product features of our roadmap.

(Openmoko Community Mailing List, February 2007)

The Openmoko project solicited enthusiastic participation from developers around the world, who wrote software, recommended changes in functionality and even self-produced viral video advertisements for the Openmoko phones. The devices that the company manufactured, including the Neo1973 and the later Openmoko FreeRunner, released in 2008, were sold with the expectation that developers would make modifications to the software, hardware and firmware. As one contributor to the publicly archived Openmoko community mailing list described, the project's application development platform created a close connection between hardware and software and placed considerable power over the functionality of the phone in the hands of individual developers:

For me the close tie in with the hardware is what interests me in the project. It is not just software. You have access to the accelerometers data, gps data, wifi, and all the open hardware. I feel that will lead to the best possible solution for the phone. It is the software working together with the hardware that truly makes the neo 1973 and openmoko such a powerful tool.

(Openmoko Community Mailing List, July 2007)

The project provided a way for developers to engage in hardware modification and for the company to use committed participants as beta testers for their projects, a strategy that was entirely new for mobile development at the time, although it was a common practice in open-source software development more broadly. One thing that make Openmoko's approach unique was its stated commitment to keeping the development process 'open' and allowing community members to observe and participate in the design process.

Tensions in Open-Sourcing

The broader forms of openness sought by the Openmoko developers posed challenges for the project leaders and the relationships they formed between their community of developers and the manufacturers of their devices. While community members wanted more access to design schematics, manufacturers were bound by nondisclosure agreements. It became increasingly evident that the structure of the mobile development process made it difficult or nearly impossible to continue to involve the community in decision-making about the device's production processes. In 2008, Sean and other core project members founders posted mailing list messages aimed at their most active developers, encouraging them to continue buying and experimenting with the Openmoko devices but not necessarily to expect that they would be able to substantively contribute to the development of a smartphone device. At the same time, other participants in the Openmoko online community had an expectation that if the were going to buy one of the smartphones, it should be a functioning consumer product, as the following list posting from 2008 illustrates:

> Third request: what *is* the warranty on the Freerunner? The warranty is essentially non-existent. It's supposedly 14 days "D.O.A.." Dead On Arrival, in its strictest definition, means that as long as the phone boots up, that's it. It doesn't matter if it's not really functional. As long as it boots up, it's technically not "DOA." If it can't make phone calls or connect to the Internet, too bad: it's not "DOA." If the GPS antenna doesn't work right, because of a hardware flaw, too bad: it's not "DOA." While I'm glad they're trying to put out a phone that's "open," I'm very disappointed in their lack of customer support. I mean, it's not some throw-away piece of crap. It's a $400 phone!
> (Openmoko Community Mailing List, August 4, 2008)

The Openmoko phone attempted to make more parts of the mobile phone open, by using open-source software, making hardware schematics open and by sharing its roadmap on the community list. Yet even among the open-source communities who participated in the development process, expectations remained that it was more important for mobile devices to operate seamlessly and be easy to use than it was important to be able to modify the phone.

The current state of the Openmoko project says a lot about the success of certain types of 'openness' and the limitations of others. After the launch of the FreeRunner in 2008, the project floundered, perhaps in part due to the incongruities between the expectations it established for developers and those it established for 'mass market' consumers—or perhaps because by this time, Apple had launched the iPhone. Compared to the Openmoko project devices, the iPhone completely lacked the open standards and libraries for application development, was based on proprietary software and, at the time, did not support any third-party applications. Yet the iPhone was an extraordinary consumer product, and as explained above, its mass market eventually permitted the development of contribution-based application stores, albeit extremely enclosed ones.

The Results of Open-Sourcing Mobiles

Despite the demise of the Openmoko project, the idea of using open-source practices to solicit application development for mobile devices has been carried forward by the Android project, which developed an open-source stack for mobile phones that was partly structured around open-source libraries and the open-source derived SDKs mentioned above. The Android project attempted to do much the same thing, except instead of creating a unique hardware project, it worked with existing manufacturers of smartphones through a trade organization called the Open Handset Alliance (OHA). The OHA contributed to the development of the Android software stack, ensuring that the software could be used on a variety of devices built by a variety of different companies, Android was acquired by Google in 2009 and Android is the operating system for roughly 50 percent of the phones used in the U.S. as of the end of 2013 (Arthur 2014).

The saga of Openmoko and the parallel rise of Apple and Android smartphones implies various ways that openness is desired, resisted or reappropriated within mobile ecosystems. Recently Openmoko has reemerged as a community project, managing the distribution of updated versions of the Neo1973 and FreeRunner prototype phones, and providing access to the source code for the Linux operating system. The goal of this project, according to its website, is "to nurture explosive innovation (such as occurred with the Personal Computer) in the field of connected mobile computing, cellphones, and ubiquitous computing" (Openmoko.org 2011). Compared with the potential opportunities for peer governance put forward by Openmoko's opening of its design roadmap, the maintenance of a platform primarily used by amateurs is less profound. Instead of openness secured by peer governance, it is contributing to a new market for open-source kits and tools, in line with similar projects that market 'open hardware' to amateurs who wish to experiment with redesigning electronic circuits (Buechley and Mako Hill 2010). This consumer-oriented openness is one of the forces behind the success of open-hardware movements, but its expanded distribution of hardware products and support of communities of practice seems largely a result of the openness of the Internet, rather than necessarily a development of its open governance across a new platform.

The Openmoko project was an attempt to introduce radical openness to the entire production process of design and manufacture of mobile devices, with results that hint at an overall shift in the mobile ecosystem away from peer governance and toward consumer consumption. This is echoed in the way that open-source processes have featured in the mobile ecosystem. Open-source contribution economies such as the application development process are clearly an important part of the current smartphone market— and indeed important for the continued utility of smartphones. Yet the economy of contribution facilitated by SDKs that provide open libraries but continue to retain the benefits of contributed app developer labor suggest that there has been a broad shift away from a framework in which developers can claim some stake in the results of their labor. This is one of the most significantly and worrying differences between the 'wired' and the 'mobile' Internet, from a perspective of labor and design.

CONCLUSION

The open Internet has some specific historical qualities that are related to the way that particular values became connected with features of its design. These created opportunities for participation in imagining the ideal frameworks for design and governance of the Internet, as well as other forms of open production including open-source software development. As everyday media practices shift to mobile platforms, individual consumers may not be aware of what they are losing in terms of openness. One of the core problems here is that we collectively share a social perception that the Internet is a technology that can support or enable individual freedom, collective participation and a range of versions of the slippery concept of 'openness.' As the practices of Internet users and the value propositions of the companies who provide connectivity to it begin to change with the adoption of mobile phones, the social practice of open governance that typified the review of Internet protocols no longer applies to the proprietary protocols used in mobile devices. The imaginary of openness does not disappear, of course—part of the influence of the FLOSS movements has been the idea that it could and should be possible to participate in the construction and governance of communication systems. What happens is that the imaginaries of openness intersect in complicated and challenging ways with the materialities of the mobile Internet. These materialities include the design of the objects but also the ways that they are embedded into particular economic, legal and social systems.

While projects such as Openmoko have attempted to address this discrepancy and bring open-source principles to the mobile realm, the promulgation of their open imaginaries is often stymied by expectations related to the materialities of mobiles. A functioning device is often more important than an open contribution framework. The ability to easily make, sell and buy apps can trump the opportunity to govern the mode of communication. The ease of use of devices like the Apple iPhone renders 'open' products clunky and pointless by comparison. Even the opportunities to transform the mobile ecology are shallow: the 'app ecology' open-sources the beginnings of the creative process and recuperates the value of the results.

What opportunities for openness remain? Are any of them as generative as the openness that developed over the 'wired' Internet? Certainly, there are disruptive practices of opening mobiles: the practice of 'modding' mobile phones by replacing their software or firmware allows some of the constraints established by the network operators to be transcended. Mods range from the very basic—'jailbreaking' a phone to allow it to work on a different network or to load applications blocked by a particular service provider—to the highly complex—replacing the memory, the antennas, the microphones, or the motion sensors in consumer devices. Modding hardware means breaking warranties and thus violating intellectual property laws. It's disruptive to the hardware industry in ways that are deeper than the challenge of a peer-production project like Openmoko, but not very well-distributed.

Perhaps it is unhelpful to expect that mobile communication platforms will recapitulate the specific openness of the Internet. Indeed, this very openness is somewhat of a historical and cultural accident, resulting from the intersection of subcultural values with the unusual expansion of a new form of communication. Yet the result has been transformative to the openness and democratic potential of our communication, in so many ways. Regardless of whether the same kinds of conditions can be met within a new communications mode, we need to develop ways and means for shared governance to apply to our communications environment. Our ability to maintain the democratic exchange of information that enriches our world depends on it.

NOTES

1. Abbate's work is the first in an evolving field of Internet histories that stress the relationships between the design of Internet systems and the social and technical imaginaries into which they are embedded. Other works in this genre include Fred Turner's *From Counterculture to Cyberculture*, which situates the emergence of both 'virtual communities' and Silicon Valley entrepreneurship in a particular positioning of the California counterculture; Gabriella Coleman's history of the political debates couched within practices of hacking and free software development; Milton Mueller's *Ruling the Root: Internet Governance and the Taming of Cyberspace*, a history and analysis of debates about how to govern the addressing system that underpins the Internet; and Zittrain's work as discussed above. There are also related works that stress

      social imaginaries more strongly, such as Robin Mansell's *Imagining the Internet*, or material structures, such as Laura de Nardis's *The Global War for Internet Governance*.

2. Together, these protocols are known as TCP/IP, and they specify how Internet data should be formatted, addressed, transmitted and routed, as well as separating the function of the Internet into several layers in order to ensure that it operates as an end-to-end network.

3. You can find the RFC Editor, which lets you view these documents, at www.rfc-editor.org.

4. In addition to the IETF and other bottom-up standards organizations the Internet Research Task Force (IRTF) and the Internet Architecture Board (IAB), Internet standards-setting bodies include the International Telecommunications Union (the ITU), which operates under a mandate to the UN, and the World Wide Web Consortium (the W3C), which brings together member organizations and the public. For more discussion of Internet standards-setting, see Brown and Marsden's *Regulating Code: Good Governance and Better Regulation in the Information Age* (Cambridge MA: MIT Press, 2013).

5. The World Summits on the Information Society were established by UN mandate, and events took place in 2003 in Switzerland and 2005 in Tunis. WSIS events concluded in 2005 with the establishment of the Internet Governance Forum, a non-decision-making forum where governments, civil society representatives and commercial representatives debate issues related to the function of the Internet. These forums are held annually, and you can view the archives as well as find out how to participate at http://intgovforum.org.

6. An SDK includes tools for technical integration of the application, as well as specifying arrangements for intellectual property and payment relating to application development.

SOURCES

Abbate, J. *Inventing the Internet*. Cambridge, MA: MIT Press, 1999.

Arthur, C. "Three Graphs to Stop Smartphone Fans Fretting about 'Market Share.'" Guardian Technology. January 9, 2014. Accessed September 9, 2014. www.theguardian.com/technology/2014/jan/09/market-share-smartphones-iphone-android-windows

Baer, W., F. Bar, Y. Hong, J. Mailland, A. Mehta and L. Movius. "Comparing Mobile Openness: Case Studies of United States, United Kingdom, France, China, India and Brazil." 39th Research Conference on Communication, Information and Internet Policy. Arlington, VA, September 23–25, 2011.

Barlow, J. P. "A Declaration of the Independence of Cyberspace." February 8, 1996. Accessed September 9, 2014. https://projects.eff.org/~barlow/Declaration-Final.html

Bauwens, Michel. "Capital and Class in Peer Production." *Capital & Class*. no. 1 (2009): 121–141.

Benkler, Y. *The Wealth of Networks: How Social Production Transforms Markets and Freedom*. New Haven: Yale University Press, 2006.

Bowker, G., and S. Star. *Sorting Things Out: Classification and Its Consequences*. Cambridge, MA: MIT Press, 1999.

Braman, S. "The Framing Years: Policy Fundamentals in the Internet Design Process." *The Information Society*. no. 5 (2011): 295–310.

Buechley, L., and B. M. Hill. "LilyPad in the Wild: How Hardware's Long Tail is Supporting New Engineering and Design Communities." Presented at *DIS Conference 2010*, Aarhus Denmark, August 16–20, 2010.

Callon, M. "Pour une sociologie des controverses technologiques." *Fundamenta Scientiae*. no. 3/4 (1981): 381–399.

Coleman, G. *Coding Freedom: The Ethics and Aesthetics of Hacking*. Princeton, NJ: Princeton University Press, 2013.

Duggan, M., and A. Smith. "Cell Internet Use 2013." Pew Internet and American Life Project." 2013. Accessed September 9, 2014. http://pewInternet.org/Reports/2013/Cell-Internet/Summary-of-Findings.aspx?view = all

Flichy, P. *The Internet Imaginaire*. Cambridge, MA: MIT Press, 2007.

Goodin, D. "Apple requires Mac App Store candidates to be sandboxed." *The Register*. November 3, 2011. Accessed September 9, 2014. www.theregister.co.uk/2011/11/03/mac_app_store_sandbox/

Held, D., and A. McGrew. "Introduction." In *The Global Transformations Reader*, 1–31. Cambridge: Polity Press, 2003.

Hofmann, J. "The Libertarian Origins of Cybercrime: Unintended Side-Effects of a Political Utopia." SSRN eLibrary. April 1, 2009. Accessed March 8, 2014. http://papers.ssrn.com/sol3/papers.cfm?abstract_id = 1710773

Indvik, L. "App Store Stats: 400 Million Accounts, 650,000 Apps." *Mashable*. June 11, 2012.Accessed September 9, 2014. http://mashable.com/2012/06/11/wwdc-2012-app-store-stats/

Isenberg, D. "The Rise of the Stupid Network." *Computer Telephony*. (1997): 16–26.

Isenberg, D., and D. Weinberger. "The Paradox of the Best Network." 2002. Accessed March 8, 2014. http://netparadox.com/

Jasanoff, S (ed.). *Ordering Knowledge, Ordering Society.States of Knowledge: The Co-Production of Science and Social Order*. New York: Routledge, 2004.

Kelty, C. "Geeks, Social Imaginaries, and Recursive Publics." *Cultural Anthropology*. no. 2 (2005): 185–214.

Levy, S. *Hackers*. New York: Doubleday, 1984.

Lipsman, A. "2013 Mobile Future in Focus." Comscore Whitepapers. February 22, 2013. Accessed September 9, 2014. www.comscore.com/Insights/Presentations_and_Whitepapers/2013/2013_Mobile_Future _in_Focus.

Mueller, M.L. *Networks and States: The Global Politics of Internet Governance*. Cambridge, MA: MIT Press, 2010.

Openmoko. "Openmoko Announces the World's First Integrated Open Source Mobile Communications Platform at Open Source in Mobile Conference in Amsterdam." 2007. Accessed March 8, 2014 (page no longer available, September 9, 2014). www.openmoko.com

Openmoko Community Mailing List. n.d. Accessed September 9, 2014. http://openmoko-public-mailinglists.1958.n2.nabble.com/

Powell, A. "Freedom Abroad, Repression at Home: The Clinton Paradox." LSE Media Policy Project Blog. 2011. Accessed September 9, 2014. http://blogs.lse.ac.uk/mediapolicyproject/2011/11/02/freedom-abroad-repression-at-home-the-clinton-now-cameron-paradox/

Powell, A., and A. Cooper. "Discourses of Net Neutrality: Comparing Advocacy and Regulatory Arguments in the US and the UK." *The Information Society*. no. 27 (2011): 311–325.

Raboy, M. "The WSIS as a Political Space in Global Media Governance." *Continuum: Journal of Media and Cultural Studies*. no. 3 (2004): 347–361.

Slack, J. "The Theory and Method of Articulation in Cultural Studies." *Social Research*. no. 3 (1997): 989–1004.

Stallman, R. "The GNU Operating System and the Open Source Revolution." In *Open Sources- Voices from the Open Source Revolution*, edited by DiBona, Ockman, and Stone. Sebastopol, CA: O'Reilly, 1999.

Star, S. L., and G. Bowker. "How to Infrastructure." In *Handbook of New Media: Social Shaping and Consequences of ICTs,* edited by L.A. Lievrouw and S.Livingstone, 151–162. London: Sage, 2002.

Stark, D. "Social Times of Network Spaces: Network Sequences and Foreign Invest-ment in Hungary." *American Journal of Sociology.* no. 5 (2006): 1367–1411.

Turner, F. *From Counterculture to Cyberculture.* Chicago: University of Chicago Press, 2006.

Wu, T. "Network Neutrality, Broadband Discrimination." *Journal of Telecommuni-cations and High Technology.* no. 1. (2003).

Zittrain, J. *The Future of the Internet and How to Stop It.* New Haven, CT: Yale University Press, 2008.

3 The Materiality of Locative Media
On the Invisible Infrastructure of Mobile Networks

Jason Farman

For several years now, I have been using locative media to "check in" to locations I visit. When I arrive to a location, I will turn on my phone, load the Foursquare app, click the "Check In" button, wait until the phone returns a list of possible locations that are near my GPS coordinates, and check in. For my own purposes, I find this practice to be a useful tool for journaling. Returning to the locations I checked in to over the past year unveils the intimate connection between space and practice, between the place and the meaning of the place, and between a location and the embodied production of that location.

One recent check-in on Foursquare stands out for me because it dramatically transformed the way that I think about locative media. On October 27, 2011, I visited a major site for the Internet on the outskirts of Washington, DC: the Equinix Data Center. This "Internet peering point" serves as one of the key places for Internet traffic on the East Coast of the United States since most of the data that moves in and out of this part of the country goes through this facility at some point. Along with most other major companies that do business in the United States, Foursquare houses their database here (run through Amazon's Cloud servers). I was able to tour the facility with my graduate students. We touched base with our tour guide on the phone that day and were asked to meet him at a nondescript door. "Since there are no signs for the facility—for security reasons—you'll pull to the end of the warehouse area and look for the only door with a handle," he said (see Figure 3.1), Andrew Blum, in his journey to uncover the material reality of the Internet, faced a similar challenge when visiting Equinix while writing his book, *Tubes: A Journey to the Center of the Internet*. He writes,

> When I showed up, I had trouble finding the door. Equinix had grown to fill six single-story buildings at the time I visited; by early 2012, four more had been added, totaling more than seven hundred thousand square feet—about the size of a twenty-story office building—all tightly arranged around a narrow parking lot. I saw no proper entrance to speak of and no signs, only blank steel doors that looked like fire exits.
> (Blum 2012, 90)

Figure 3.1   The door that led to the Equinix Internet Peering Point on the outskirts of Washington, DC. Image © 2011 Jason Farman.

He goes on to note that, according to David Morgan, the director of operations at this site, this kind of confusion is the goal of the design of Equinix: "customers are reassured by the anonymity of the place" (Blum 2012, 90).

The "door to the Internet" led to a room that was only as big as a medium-sized elevator. Brian, our tour guide, placed his hand on a biometric scanner and punched in his passcode in order to open the doors into the next room. This led us to the security area, where we showed our identification and were given visitor badges. Brian scanned his hand on another biometric scanner, punched in his passcode, and led us to the waiting area, which had a conference room where we were told about the history and current practices of this Equinix facility. He led us to the next set of doors, which was the entrance to the data center. Once he scanned his hand and entered his passcode one last time, we walked into a frigid, dark and extremely loud environment. Here, there were long stretches of aisles in all directions, bounded by black steel cages housing servers stacked to the ceiling (see Figure 3.2). These servers run nonstop, twenty-four hours a day, and produce so much heat that one of the primary concerns for these facilities is keeping the electricity running to the air conditioning system that keeps everything cool.[1] Thus, upon entering the data center, I could barely hear Brian talk above the noise of cold air blasting down onto the front faces of the server racks. Above my head, running down the center of each aisle, was a yellow tray that held the fiber-optic cables running to and from each server. Along these cables, data was flowing around the world at speeds so incredibly fast that it was difficult for me to comprehend.

It was here that I pulled out my mobile phone, loaded up Foursquare, and checked in to the Equinix data center over a 3G network. Knowing that

Figure 3.2   The inside of the Equinix Internet Peering Point in Ashburn, Virginia. Behind the cages are the servers that are run by major companies like Amazon, EA Games, Verizon and Google. The yellow trays along the aisles hold the fiber-optic cables running between these server cages. © 2011 Equinix, Inc. All rights reserved. The Equinix logo is a trademark of Equinix, Inc.

Foursquare's databases were housed somewhere at this facility—located on servers behind one of these cages—I began to wonder about the flows that my information took. What information pathways were necessary in order for me to be "located" and to broadcast my location to my network of friends on Foursquare? What does the infrastructure behind locative media and mobility actually look like?

Tracing the flows of my locative data turned out to be an enlightening endeavor. Combining research on the ways that the mobile Internet works with physically walking around and locating cell towers and antennas, I discovered that as my mobile device sent out its signal from the Equinix building, it connected (through a line-of-sight link) with my carrier's nearest cell antenna through what is termed an "air interface." Since I was in an area with enough coverage, my signal connected directly to a cell tower that was linked in with the technology that routes my request through the fiber-optic (or sometimes copper) cables that connect to the Internet (called a "backhaul"). As soon as my signal connected with this cell tower, the rest of the pathway was not wireless; instead, the journey to Foursquare's database took place entirely through the material, hard-wired, tangible infrastructure of the mobile Internet. The signal ran down the wires of the cell tower (see Figure 3.3), connected into the fiber optics of the Internet

Figure 3.3   The base of the cell tower that my phone connected with near the Equinix Peering Point, showing the wired infrastructure that my data flowed through. Image © 2013 Jason Farman.

infrastructure, and took various routes from the cell tower to the larger cellular network that directs the data packets that are sent from a phone. After being directed by my carrier's network, the packets of data sent from my phone came shooting right back to my location—the Equinix data center—in order to access Foursquare's database. It is quite possible that the request made by my mobile device traveled on the fiber-optic cables directly over my head in the data center where I was standing as this request went down into one of the cages and pulled the relevant data from the server. Here, using the GPS coordinates delivered to the server by my mobile device, Foursquare was able to pull up a list of nearby locations. It then sent that information back along the fiber-optic cables, back out to the cellular network, up the cell tower lines, and back out to the antennas, eventually landing right back on my mobile device.

When tracing the flow of my data, it struck me how circuitous the pathway was to send and receive information. Even more striking was the fact that much of the journey of my data took place across a very static/fixed infrastructure. In fact, much of what we consider to be "mobile" media is generated through very nonmobile technologies such as the cell tower and fiber-optic cable.

The largest take-away for me gained by tracing the flows of my mobile and locative data—with which this chapter will be most concerned—is that most of these interactions took place at a level that was far beyond my awareness. I tend to imagine that locative media begin and end at the level of the interface; however, this sentiment couldn't be further from the truth. The vast majority of locative media takes place well beneath the level of the interface, with technologies communicating with technologies in ways that exceed my own sense perceptions. Therefore, as we continue to theorize location-aware technologies, especially as they impact the practices of embodied space, social interaction and site-specificity, we need to look beyond the human-to-human connectivity to take into consideration both the human-to-technology and the vital technology-to-technology interactions that take place beyond the realm of the perceptible. Ultimately, what is gained from such an inquiry is a necessary insertion of objects into our understanding of the production of space and of the ability for embodied people to truly engage a political practice of difference. My sense of embodied identity is not simply gained through social interactions with other humans; instead, it is continually constituted by what I term a "sensory-inscribed" engagement with people, objects, social and cultural structures, protocols and the spaces produced through the interactions among these things.

Ultimately, to argue for this mode of embodiment, it becomes apparent that many of these interactions are not visible to us. They tend to recede into the background of our practices with locative technologies. In fact, many of these technologies are designed to withdraw from view (if they are designed well, according to the designers). Thus, as I try to unveil the role that objects play in the practice of location-aware technologies, my hope is to push on the categories of "visible" and "invisible" to trouble the ways that these concepts are implemented in the design and everyday practices of mobile and locative media.

THE AUDIENCE OF LOCATIVE MEDIA

When I checked in to the Equinix facility, I attached a brief note to my check-in saying, "Here to 'see the internet' with my grad students" (see Figure 3.4). This note, along with my location, was broadcast to those in my network on Foursquare. So, if any of my seventy-eight friends and

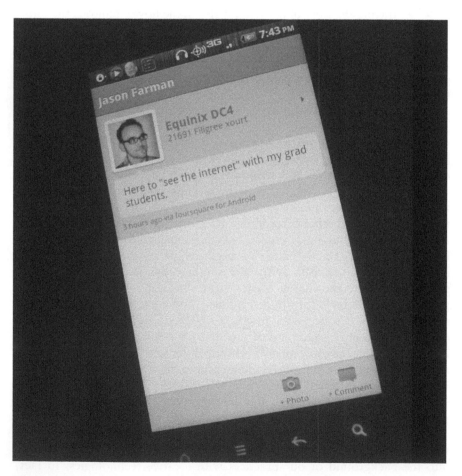

Figure 3.4   An image of my check-in to Equinix on Foursquare, which broadcasted my location, activity and companions to those in my network. Image © 2011 Jason Farman.

colleagues happened to load Foursquare that evening—or if any of them had push notifications enabled on their devices that would automatically send my message to the screens of their mobile devices—they would have known that I was visiting this site with my small group of graduate students from the University of Maryland. This seems to be the most obvious audience for locative media and, in fact, most scholarship on locative media tends to focus on it as an extension of social media sites like Facebook by bringing it out onto the streets and into the everyday movements of these networks of people. For me, as I mentioned at the beginning of this chapter, my primary use of Foursquare is as a kind of spatial journal to log my movements throughout a given day. So, my network was not my primary audience; instead, I thought of myself as the audience of this message, which I received exactly one year later on email in a message that read: "Your check-ins from one year ago today." This service, now called Time Hop (though once cleverly called 4SquareAndSevenYearsAgo), sends reminders to users of Foursquare (and other services like Facebook and Twitter) of what they were doing a year ago on this day.

Yet, my check-in to Equinix forced me to ask: who is the primary audience of locative media? Could it possibly be the objects that are interacting and exchanging information with each other beyond the realm of my human awareness? When I loaded my phone to check in, I was searching for a 3G signal and searching for my location from Foursquare's database. Simultaneously, my phone was searching for the nearest antenna and backhaul, which searched for the correct IP address in order to connect to right fiber-optic line leading down the long yellow tray that ran above my head and down into a cage in the data center, connecting to the exact server that held the information that corresponded to my GPS coordinates. Here, machines are talking to machines. There is an entire network of connectivity that encircles this one act of "checking in," and I am just one node in that network, and while I might have been the initiator of this particular request ("Foursquare, please locate me and log my visit to this location"), my phone was already engaging this network of technological objects long before I made the request. My phone is able to navigate handoffs between various network "cells"—hopping from antenna to antenna to maintain a seamless connection so I never have to drop a call or receive an error message on a website I'm browsing. My phone is also tracking my movements, often without me giving it permission to do so (which has, in recent years, caused major controversies). It connects with GPS satellites—those celestial objects constantly orbiting the earth and only visible to me at night when they look like extremely slow shooting stars—three at a time in order to triangulate my position. My phone runs constant checks on my email, on my text messages, my voicemails, messages to me on Twitter, and a host of other services I have set up to run in the background while I go about my day. This is all happening right now as my phone sits at the bottom of my pocket.

Such interactions broaden our understanding of what constitutes an "audience" and extends Alice Rayner's theorization of what an audience is for a performance studies perspective. She writes that the term audience is "a model for intersubjective relations as opposed to a model for a unified community; to view the audience, that is, as a 'boundary condition' in the act of understanding another and, as a result, of understanding the constitution and contradictions of its own differences" (Rayner 1993, 6). The audience is not a coherent, unified entity based on collective experience and empathy; instead, the audience exists as a practice of *imagined collectivity* between listeners that simultaneously offer feedback, between those in a constantly evolving context that have affective/effective impact on/by the messages being sent and received.

When this begins to truly sink in, I realize how small my Foursquare request was and that I am not the primary audience for locative media. I am only *one* audience member among many, including my phone, the cell antennas, the cell tower, the fiber-optic cables, the servers, data talking to data, machines talking to machines, signals communicating to infrastructure. There are a whole host of communicative objects that are connecting and serve as the "audience" of locative data. This shifts the entire focus of locative media in fairly radical ways. Instead of being this solipsistic interface for the oversharing of banal information (e.g., "To all of my friends: I am currently at the corner gas station filling up my car."), locative media highlight the site-specific interactions among human agents and the informational objects that are weaving themselves into the fabric of our everyday lives.

MOBILE MEDIA AND OBJECT-ORIENTED PHENOMENOLOGIES

The role these informational objects play in the practice of locative media initially rubbed against the grain of my theorizations of embodied space in a mobile media age. As I worked to define "interface" in an early draft of my book, *Mobile Interface Theory: Embodied Space and Locative Media*, I wrote that the interface is "constituted as a larger set of social relations" and is the nexus of these relationships. Therefore, the mobile phone, in and of itself, is not an interface (I still hold this to be true). Only when the mobile phone serves as the nexus of relationships and interactions does it become an interface. I draw here largely from Johanna Drucker's arguments when she writes, "What is an interface? If we think of interface as a *thing*, an entity, a fixed or determined structure that supports certain activities, it tends to reify in the same way a book does in traditional description. But we know that a codex book is not a *thing* but a structured set of codes that support or provoke an interpretation that is itself performative" (Drucker 2011, 8). However, my attempts to think beyond the "thingness" of the mobile-phone-as-interface (and instead think of the interface as the nexus of relationships) led me down an ill-reasoned path. In this early draft of my book, I went on to say, "The mobile device, on its own, is nothing but a beautifully designed paperweight." For me, as I worked to theorize embodied space as something that is produced through *practice*, the mobile phone sitting on my desk seemed to not engage these practices because I was not directly interacting with it. Once I began to see beyond this narrow box of human-computer interaction and practice, the device's engagement with its own practices, its own audiences, its own content opened up an entire realm vital to my theorizations of the production of space.

My theorization of embodiment is founded on the ideas in phenomenology, especially the work of Maurice Merleau-Ponty. This mode of inquiry prioritizes an experience of the world gained through the body and through the senses. It takes up experiences, sense perception, the body's role in knowledge production and the philosophies of being as its main objects of study. It is focused on subjectivity and embraces the limited perspectives of human perception as the main window through which we understand the world. By inserting mobile objects—the materiality that serves as the foundation of locative media—into a phenomenological approach to understanding embodied interaction, we begin to see some of possible limits to phenomenology (limits, I will argue, that don't accurately characterize the phenomenological approach of those like Merleau-Ponty and Martin Heidegger).

In a 1946 presentation of his work to the *Société francaise de philosophie*, Merleau-Ponty was challenged to defend his work against the idea that phenomenology was solipsistic and always must come back to the individual's perspective on the world. During the question and answer session of his presentation, another philosopher of the time, Emile Bréhier, lobbed a complaint against Merleau-Ponty and his phenomenological approach. His complaint is one that would thereafter be echoed multiple times by other detractors of phenomenology: "When you speak of the perception of the other, this other does not even exist, according to you, except in relation to us and in his relations with us. This is not the other as I perceive him immediately; it certainly is not an ethical other; it is not this person who suffices to himself. It is someone I posit outside myself at the same time I posit objects" (Merleau-Ponty 1964, 28).

Indeed, phenomenology, as with the main object of study for this chapter—locative media—can be seen as giving an overemphasis to the individual rather than the community or others that are able to exist beyond our realm of understanding. This kind of "immanence" of the individual comes at the cost for any existence of "transcendence" of the other. As Jack Reynolds puts it, referencing Emmanuel Levinas's concern with this approach to being in the world, "phenomenology hence ensures that the other can be considered only on the condition of surrendering his or her difference" (Reynolds 2004, 125).

Michael Yeo phrases the problem of immanence and transcendence by asking, "How is it possible to experience the other as other—as really transcendent—given that I cannot but experience her in relation to my own immanent frame of reference?" (1992, 38). However, any theorization of difference requires the ability for us to engage transcendence, for a recognition that not all things exist on the plane of immanence, but instead can exist beyond our understanding and realm of experience. People need to be able to exceed their individual understandings and frames of reference in order for there to be any sense of knowledge, surprise and cross-cultural exchange. As Reynolds argues, "Not only can interactions with the other involve us in a renewed appreciation of their alterity (i.e., the ways in which they elude us), but the other is equally importantly that which allows us to surprise ourselves, and move beyond the various horizons and expectations that govern our daily lives" (2004, 128). Within this framework of alterity, we *must* include objects. Objects have to be considered as often exceeding our realm of understanding (and thus existing on the horizon of transcendence).

The flow of my mobile data offers an important example of this: most people, including myself prior to this investigation, typically engage the mobile Internet without any sense of awareness about the larger network of objects that make such connectivity possible. Though these objects and their interactions with one another often exceed our realm of human perception, they nonetheless shape the sense of self via location-aware technologies. We still deeply understand our sense of space even though we may not be aware of the informational objects that help produce that sense of space. For a *sensory-inscribed* phenomenology of embodiment, which stakes a claim in the continued value of phenomenology for analyzing our sense of being in the world, transcendence is always at the forefront of what it means to be human in a technological world. This transcendence, however, takes place on many phenomenological levels. The phenomenology of sensory-inscription, as produced simultaneously through the senses and through various cultural inscriptions, takes into account the distinction that cognitive scientists have made between *cognitive awareness* and the *cognitive unconscious*. There are things that we gather through human sense perception that we're aware of: I turn on my phone, load Foursquare, compare its results to where I think I am, check in and finally compare my position in the world with my friends and colleagues. These are actions of which I am conscious. My awareness of these various levels of mediation involved in the ways space is produced ultimately results in an embodied sense of proprioception (i.e., of knowing where my body fits as it moves out into the world, of knowing the spatial relationships that constitute my place in this locale). However, there is much beneath the surface of human perception that is still vital to this sensory-inscribed experience of space. This takes place at the level of the "cognitive unconscious." Many of the aspects of life and the uses of our emerging technologies require our ability to either filter out excess information in order to focus or, alternatively, to simply not be aware of certain aspects of our lives and surroundings (both biologically, such as not knowing how many times you've blinked in the last five minutes, and socially, such as not knowing the name of the person at the Foxconn factory who put the screen on your iPhone).

While much of our experience of location-awareness through our technologies does engage the cognitive unconscious, it is often *designed to do so*. For many designers of mobile devices and mobile applications, a good design is one that is not noticeable. If done well, it should recede into the background of your everyday life. Designers dating back to at least Mark Weiser's work on ubiquitous computing have sought to design for invisibility or what Jay David Bolter and Richard Grusin have called the immediacy of the interfaceless-interface (Bolter and Grusin 1999, 23). As Weiser argues, "The most profound technologies are those that disappear. They weave themselves into the fabric of everyday life until they are indistinguishable from it" (Weiser 1991, 66). This is what Heidegger calls "readiness-to-hand," and what philosophers in the field of Object-Oriented Ontology (OOO) like Graham Harman term "tool-being."[2] The tool, in its essence as a tool (its tool-being), is invisible to us. Only when the tool stops working (and, for Harman, no longer exists in its essence as the tool it once was) do we actually notice the tool. This move from ready-to-hand to present-at hand is the move from invisibility to visibility.

MOVING MOBILE MEDIA FROM THE REALM OF THE INVISIBLE TO THE VISIBLE

The categories of "ready-to-hand" and "present-at-hand" are rarely clear cut and distinct from one another. Emerging mobile devices are representative of the bleed-over between these categories. While designers and theorists of ubiquitous computing, like Weiser, hope to achieve a design that so intimately weaves itself into the fabric of our everyday lives that we rarely notice it as a distinct interface from "the interface of everyday life" (Farman 2012, 86–87), there is also a cultural desire for our devices to engage in this kind of seamless integration *and* be a visible object of significance. The device's immediacy should be obtained while allowing for the device to still function as a status symbol, a technological fetish object and a signifier of a person's brand loyalties. The fluctuation between these levels of invisibility and visibility are a result of the power structures (such as capitalism) that are invested in maintaining various levels of visibility and invisibility and the cultural desires that are founded on such structures.

For mobile and locative media, an analysis of the relationship between visibility and invisibility is a productive one, and one of the best examples is seen in the attempts to camouflage cell phone towers (and their resulting hypervisibility). I see this daily as I drive to work, making the thirty-five-minute commute to College Park, Maryland. As I drive through the northern part of the town of Silver Spring, at the corner of Bonifant Road

Figure 3.5   A cell tower disguised as an evergreen tree in Silver Spring, Maryland. Image © 2012 Jason Farman.

and Layhill Road, there is a brick building with cell antennas perched on the top. These antennas are covered with a faux-brick veneer in order to make them blend in with the building. Driving another few minutes east on Bonifant, there is an extremely large tree standing in a field, twice as large as any other tree in the area. Even from a distance, it is apparent that this is a cell tower disguised to look like an evergreen tree (Figure 3.5).[3] As I turn south onto New Hampshire, I see an interesting contrast between a huge cell tower on the busy intersection of Randolf Road and New Hampshire that stands high above the ground without any attempt to disguise it, and a cell tower further down New Hampshire, which blends into a church steeple near Adelphi Road.

This attempt to make the infrastructure of mobile technologies invisible is, in large part, one of the only recourses a community has as a response to the excessive visibility of these towers. As Ted Kane and Rick Miller note, the Telecommunications Act of 1996 essentially removed local oversight about where cell towers could be placed, thus "local governments are limited to regulation based on community planning standards, largely imposing some form of *visual control*" (Kane and Miller 2008, 152).

These methods of visual control, however, almost always make the infrastructure more pronounced and *more visible*. The attempts at disguise only draw our eye to the towers rather than allowing them to blend in. In thinking of an object-oriented approach to the study of these towers, this invisibility-made-visible led me to ask, "Who do these attempts at invisibility address?" Since mobile technologies work on line of sight (the signal from your mobile phone must connect with an antenna based on a direct, visible connection) the height of the evergreen cell tower in Silver Spring, Maryland, is designed to be visible to mobile devices in the area. The technologies work because they are visible to one another. However, the invisibility of the tower—the attempt to camouflage it as a tree—is done for the human users and inhabitants of this region.

While those addressed by this visible/invisible divide might seem obvious, the political consequences are significant. As Lisa Parks notes, the attempts to make our technologies invisible, as profoundly seen in the attempts at disguising cell towers, highlight a cultural desire to remove particular technological objects from view and, thus, to remove the political consequences of having those objects play a vital role in the ways that we practice space, identity and community creation. She writes,

> Perhaps the ultimate irony of the antenna tree is that it actually exposes more than it hides and in this sense can be thought of as a site for generating further public knowledge about the materiality of wireless and other network systems. We are socialized to know so little about the infrastructures that surround us, even though many of us use mobile phones each day. Would our experience of mobile telephony change if we knew more about the architectures of signal distribution? It is difficult to say, but we certainly would have a different relation to the technology if we understood it as something more elaborate and expansive than something that rings in our purse or vibrates in our pocket.
>
> (Parks 2009, n.p.)

The move to make our mobile objects and infrastructures invisible is to deny the "vibrant matter" of things and the essential part they play in the ways that we think about being human in this pervasive computing age. This resonates with Jane Bennett's concerns when she writes, "How would political responses to public problems change were we to take seriously the vitality of (nonhuman) bodies? By 'vitality' I mean the capacity of things— edibles, commodities, storms, metals—not only to impede or block the will and designs of humans but also to act as quasi agents or forces with trajectories, propensities, or tendencies of their own" (Bennett 2010, viii). She goes on to argue that things like patterns of consumption would change "if we faced not litter, rubbish, trash, or 'the recycling,' but an accumulating pile of lively and potentially dangerous matter" (Bennett 2010, viii).

What would be the consequences of approaching mobile infrastructure in a similar way? Would advocating for a "vibrant materiality" of our mobile devices and infrastructures have an impact on things like consumption and purchasing patterns? On the ways we understand and legislate labor in the digital age? By inserting these vibrant objects into our understanding of locative experiences through mobile media, our phenomenological engagement with the world requires us to address the need for otherness and alterity: not only is locative media about connecting with one another, but it is about connecting with the various information objects around us, the spaces in which those objects exist, the labor that produces these devices and infrastructures and, ultimately, about objects connecting with objects. Objects like the cell tower and the data center are extreme versions of otherness, communicating and interacting with each other in ways that far exceed my realm of understanding. Yet, the ways we make these things invisible very likely is derived not from a stance of seeking the transcendence of others, but instead from a stance that seeks immanence of the self, that seeks to place the individual far above any of the networked social connections that are responsible for the constitution of the self.

CONCLUSION: TOWARD VISIBILITY

Realizing the various objects that make something as simple as a locative application like Foursquare work has been illuminating for me. I see my mobile phone differently. I see the landscape around me differently, constantly noticing cell towers, antennas and networked infrastructures of all kinds, and while most of these things still remain outside of the realm of full visibility—for some things, like data transfer speeds across a fiber-optic cable or the signal emitting from my mobile device, I will never be able to fully grasp on a sensory level—I believe the push toward increased levels of visibility with our mobile tools and infrastructure is ultimately vital to our understanding of identity and difference broadly. The push to make our objects disappear ends up mirroring our push to make otherness disappear, and, ultimately, as we continue to engage in cultural analysis and theory, this kind of difference plays a central role. As those modes of inquiry continue to grow, they will need to include objects as a central object of study. For through the study of objects like emerging locative and pervasive computing tools, we can begin to locate the essential relationship between humans and objects, and between visibility and alterity.

NOTES

A shorter version of this chapter first appeared in *The Routledge Handbook of Mobilities,* edited by Peter Adey, David Bissell, Kevin Hannam, Peter Merriman and Mimi Sheller (New York: Routledge, 2014), 233–242.

1. Brian noted that in the event of a major power outage (e.g., in a natural disaster or a "terrorist attack"), the gasoline to keep generators running would first be rationed out to area hospitals; second in line would be Equinix. The need to

keep the Internet running is prioritized as a central to national safety in such circumstances.

2. See Graham Harman, *Tool-Being: Heidegger and the Metaphysics of Objects* (Chicago: Open Court, 2002). Harman, as well as Ian Bogost in his book *Alien Phenomenology, Or, What it's Like to be a Thing* (Minneapolis: University of Minnesota Press, 2012) argue for an "object-oriented ontology" that extends the theorizations of those like Bruno Latour and actor-network theory by arguing for "things" that can be theorized even outside of the network. Bogost's approach finds limitations in actor-network theory since "entities are de-emphasized in favor of their couplings and decouplings."

3. Interestingly, since it is "evergreen" and surrounded by mostly deciduous trees, there are a remarkable number of birds living in this cell-tower-made-tree because it offers year-round shelter.

SOURCES

Bennett, Jane. *Vibrant matter: A Political Ecology of Things.* Durham: Duke University Press, 2010.

Blum, Andrew. *Tubes: A Journey to the Center of the Internet.* New York: Ecco, 2012.

Bolter, David, and Richard Grusin. *Remediation.* Cambridge, MA: MIT Press, 1999.

Drucker, Johanna. "Humanities Approaches to Interface Theory." *Culture Machine.* no. 12 (2011): 1–20.

Farman, Jason. *Mobile interface theory: Embodied space and locative media.* New York: Routledge, 2012.

Kane, Ted, and Rick Miller. "Cell Structure: Mobile Phones." In *The Infrastructural City: Networked Ecologies in Los Angeles,* edited by Kazys Varnelis. New York: Actar, 2008.

Kihlstrom, John F. "The Cognitive Unconscious." *Science.* no. 237 (1987): 1445–1453.

Merleau-Ponty, Maurice. *The Primacy of Perception.* Evanston: Northwestern University Press, 1964.

Parks, Lisa. "Around the Antenna Tree: The Politics of Infrastructural Visibility." *Flow.* no. 8 (2009). Accessed February 1, 2014. http://flowtv.org/2009/03/around-the-antenna-tree-the-politics-of-infrastructural-visibilitylisa-parks-uc-santa-barbara/

Rayner, Alice. "The Audience: Subjectivity, Community and the Ethics of Listening." *Journal of Dramatic Theory and Criticism.* no. 2 (1993): 3–24.

Reynolds, Jack. *Merleau-Ponty and Derrida.* Athens: Ohio University Press, 2004.

Weiser, Mark. "The Computer for the 21st Century." *Scientific American* (September 1991): 66–75.

Yeo, Michael. "Perceiving/Reading the Other: Ethical Dimensions." In *Merleau-Ponty, Hermenutics, and Postmodernism,* edited by Thomas W. Busch and Shaun Gallagher, 37–52. Albany: State University of New York Press, 1992.

4 Labors of Mobility

Communicative Capitalism and the Smartphone Cybertariat

Enda Brophy and Greig de Peuter

THE MOBILE CIRCUIT OF EXPLOITATION

"The world is awash in low-cost smartphones," says one analyst of the current state of the global market for mobile devices (Restivo in El Akkad and Marlow 2013). As mobile communication technologies become increasingly ubiquitous and profit margins for manufacturers begin to narrow, so-called emerging economies are singled out as a new front in the "smartphone's global price war," as a *Globe and Mail* headline put it. The unprecedented reach of handheld connectivity is the background to one BBC business reporter's claim that 2013 was "the year we all went mobile" (Wall 2013). Previously bullish predictions regarding the growth trends that would characterize mobile media, the BBC writer noted, had been entirely verified: "More cloud computing, services and storage. More mobile devices with more apps, more games, more browsing, more shopping and more sharing with our 'friends' on social media. All of which increases the demand for bandwidth, faster speed and universal coverage" (Prentice in Wall 2013).

Such commentary points to the ascendancy of what Jodi Dean (2009, 17) terms "communicative capitalism," a twenty-first century order that is increasingly dependent upon the proliferation of "communicative access and opportunity." Massive corporate investment in privatized telecommunications infrastructure throughout much of the world during the 1990s has inaugurated a situation in which communicative scarcity has "begun to recede" (Schiller 2007, 81). Rather than outright exclusion, the differential inclusion of populations as consumers of information technology services (Qiu 2009) has become a norm in a post-Fordist economy that extracts financial value from communication, information and knowledge.

Hardly a celebration of ubiquitous connectivity, Dean's concept of communicative capitalism suggests that the profusion of communication has not enhanced equity or deepened democracy, as earlier generations of media scholars on the left might have anticipated. This process has, on the contrary, fueled a networked economy that is the scene of familiar class asymmetries and consolidating corporate powers. One way that the persistence of inequality in the information age can be made visible is through an examination of labor, a facet of social life that has been identified as a long-standing "blind spot" within studies of communication (McKercher and Mosco 2006, 493).

If mobile communication is, as Manuel Castells (2009, 62) wrote, the "fastest diffusing communication technology in history," this chapter sets out to illuminate some of the manifold labors sustaining—and transformed by—this diffusion. In this task we take a cue from labor researcher Ursula Huws (2003), who introduced the figure of the "cybertariat" in an analysis of the reorganization of work through information and communication technologies. Against nineties-era New Economy narratives that proclaimed the upward spiral of worker skill, the smoothing of global space, and the weightlessness of commerce, with the category of the cybertariat, Huws advocated a critical project to "make visible the material components of this virtual world," one that would address "human beings, in all their rounded, messy, vulnerable materiality" (Huws 2003, 127, 151).

This chapter adopts such a materialist approach for the study of mobile communications—one that surveys the typically invisible, vastly diverse and globally interlinked forms of labor set in motion by the burgeoning wireless sector. Traced out is a circuitry rarely examined in mobile communication studies, namely, the circuit of exploitation constituting one of communicative capitalism's ideal commodities, the smartphone. Of the myriad labors of mobility animating the turbulent cycle of wireless accumulation, we touch on just six moments: *extraction*, where the earthly material necessary for mobile networks is mined; *assembly*, where the mobile device and the components inside it are manufactured within complex supply chains; *design*, where the applications that attract mobile subscribers are conceived and engineered by high-tech developers; *mobile work*, where the wireless handset functions as a platform for the performance of labor as such, both paid and unpaid; *support*, where mobile consumers are mediated at a distance by call center staff; and, finally, *disassembly*, where the ostensibly obsolete mobile device is dissected by e-waste workers for residual value, spinning off a secondary cycle of accumulation revolving around molten metal.

Mapping the assemblage of toil upon which mobile communication is built, some readers might react, is an exercise in stating the obvious. To such a criticism we reply that the banalization of exploitation is one of communicative capitalism's greatest achievements. So while this chapter is an attempt to confront the marketing imaginaries of endless communicative expansion surrounding the mobile Internet with their profoundly material conditions of possibility, we also emphasize that the smartphone cybertariat is not identical to the circuit of exploitation. Punctuating this fragmentary portrait are hotspots where the living labors of mobility variously refuse, challenge or escape the frame within which communicative capital seeks to seclude them. The point, then, is not simply to trace out a circuit of exploitation. Our foray into the labors of mobility has also sought moments of resistance. What remains to be seen is whether a wider labor movement might once more claim the utopian impulse of social connectivity as *its* mobilizing imaginary.

EXTRACTION

The initial moment on the circuit of exploitation that we trace out is that of extraction. A traditional materialist definition of labor as the transformation of nature loses nothing of its relevance at the frontiers of communicative capitalism. Sparkling retail stores displaying the smartphone belie this commodity's stubborn dependence on raw materials dug from dirt. Confronted in this first moment is the literal bedrock of the mobile device, its polished components containing resources such as gold, tungsten, tin and, most controversially, tantalum, a silicate derived from columbite-tantalite, or coltan.

Even as the mobile Internet smoothes communication for some, the case of coltan has brutally illustrated the enduring significance, scope and scourge of "geographic difference" (Harvey 2010, 149). Of the territories endowed with coltan—Australia, Brazil and Canada among them—the vastest known deposits lie in the Democratic Republic of Congo (Marlow and El Akkad 2010), an African nation whose mineral wealth is grossly juxtaposed to its rank in the 186th slot on the United Nations' Human Development Index in 2012 (United Nations 2013). Tantalum, an element vital to capacitors, is used in countless electronic gadgets. In the early 2000s, however, "mobile phones" were singled out as having become the "main attraction" to coltan in the DRC (Pole Institute 2002, 5).

Mining in this former Belgian colony became increasingly informalized from the 1970s onward (International Alert 2010, 15). Rife with corruption, the dictatorial regime ruling what was then Zaire instructed its subjects, "fend for yourselves" (cited in Mantz 2008, 44). Big mining companies abandoned Congolese sites amid Africa's World War (1998–2003). A notorious early-twenty-first century display of global capitalist polarity, feverish demand in the overdeveloped world for game consoles, mobile phones and other consumer electronics fed a brief coltan boom, adding incentive for competing regional militias to set up mining operations—their productivity often arising from the "barrel of a gun"—so to bankroll their contribution to an internecine conflict taking more than three million lives (Smith and Mantz 2006, 77). That war ostensibly over, one observer cautioned, "When the spider leaves, the web remains!" (Holt in International Alert 2010, 36).

While information-at-your-fingertips may be a familiar refrain in a mobile Internet milieu, anthropologists James Smith and Jeffrey Mantz (2006, 77) report that in their fieldwork in a DRC mining town they were hard-pressed to meet people aware of coltan's eventual application. Although "[n]ature does most of the work" where the development of raw materials is concerned (Marx 1857, 179), setting their value in motion turns on the "mining proletariat" transforming them (Engels 1845). Human rights groups began documenting coltan's labor economy (International Alert 2010; Pöyhönen et al. 2010; Pole Institute 2002). The picture that emerged was of primarily small-scale, artisanal and fleeting mines, dispersed across remote forests, where diggers—typically migrant, young and male—use rudimentary tools to variously strip land, burrow underground, crush ore or haul coltan—jobs whose growth has coincided with deteriorating agricultural skill.

Laboring in a context where sexual violence and ecological degradation are rife (International Alert 2010), this precarious mining proletariat inhabits an informal economy whose other actors have spanned from militias imposing illicit taxes on distributors, to regional exporters and smugglers, through to foreign traders. The majority of DRC coltan is reported to be shipped to Chinese smelters, after which it is inserted into the component manufacture process in an often opaque smartphone supply chain (Global Witness 2010, 15; Ma 2010, 3). The world market in tantalum-filled capacitors has an estimated worth of $2 billion (Marlow and El Akkad 2010). Diggers, meanwhile, can eke out a few dollars a day, a barely subsistent livelihood marring neoliberal visions of "mobile-led economic development" ("Mobile Marvels" 2009).

With slogans like "No blood on my mobile," NGO efforts to malign big-brand buyers of DRC coltan augured the passage in 2010 of U.S. legislation requiring corporate disclosure of mineral sources as part of the Dodd-Frank Wall Street Reform and Consumer Protection Act. Channels through which coltan circulates are, however, labyrinthine, and frequently clandestine; given the paucity of sources one journalist wagered, "there is no such thing as a 'conflict-free' phone" (York 2010). For the DRC's part, a six-month mining ban was implemented in 2010 in the provinces where coltan is concentrated; rooting out mafia-controlled mining was the stated goal, though skeptics worried that this intervention risked pushing mining into even more insecure patterns and failed to stop profiteering by state military commanders themselves (Doya 2011; Hogg 2011).

Apple (2014, 16) claimed that as of 2014 the "tantalum smelters in Apple's supply chain were verified as conflict-free by third-party auditors. . . ." While the auditing process used in this case is itself opaque, the advocacy efforts of human rights organizations have certainly pressured smartphone makers to inch toward greater supply chain transparency, whether legislated or voluntary, the efficacy of which remains to be seen. Here we simply emphasize two points from the case of coltan that are integral to a materialist approach to mobile communication: first, that the wireless handset has been a catalyst of profoundly material transformations in the Global South, where imaginaries of expanding access coexist with realities of environmental racism; and, second, as Mantz (2008, 44) has so forcefully conveyed, that "we would not be speaking about the great promises that inexpensive digital products have for bridging communicative boundaries the world over" were it not for the background work of scores of Congolese performing grueling—but for that no less ingenious—manual labor in a setting shaped by extreme violence. This is the raw materiality of communicative capitalism.

ASSEMBLY

Despite its apparent "dematerialization," capitalism's prevailing tendency is, Huws (2003, 131) has argued, the forward march of commodification through the mass production of tangible goods. The mobile Internet is a straightforward instance: mobile services form the attraction; accessing those services requires a physical mobile device. From smartphones to no-frills cells, more than one and a half billion handsets shipped in 2011 (IDC 2012)—the astonishingly productive output of the dexterous labor animating the second moment on the circuit of exploitation, assembly. This rote work is commonly outsourced or offshored by the biggest brands dominating the mobile industry (Wilde and de Haan 2006, 19). Much of it occurs in factories in the Global South in Special Economic Zones (SEZs), neoliberal territories par excellence where communicative capitalism's ongoing reliance on mass aggregation of industrial labor is manifest.

Nowhere is this more obvious than China. Of all the mobiles that were sold globally in 2012, it is estimated that half were made in that country (Custer 2013). Byzantine supply chains have generally made it difficult to locate assembly labor with precision. A glaring exception is the multinational corporation known as "the world's biggest contract maker of mobile phones" (Lim and Culpan 2010): Taiwan-based Foxconn catapulted out of relative obscurity and onto the cover of *Wired* magazine (Johnson 2011) when in 2010 news spread of suicides among its young, and majority female, workforce that snaps iPhone and iPad parts into place for Apple (Barboza 2010; Wagner 2010). Other clients have included Motorola, Nokia and Sony Ericsson. With 1.2 million employees, Foxconn's parent corporation, Hon Hai Precision Industry, is the world's tenth-largest employer (Alexander 2012). The greatest share of Foxconn's labor pool is in mainland China, with about four hundred thousand employees in Shenzhen, a premiere Chinese SEZ (Chang 2013).

Foxconn's notoriously fortified Shenzhen facility was spotlit in a report by Hong Kong's Students and Scholars Against Corporate Misbehaviour (2010). Informed by undercover jobs and employee interviews, this exposé cataloged grievances that ran the gamut: wages that fall generously below the cost of living—explaining workers' willingness to do overtime whose legal limit is habitually violated; a hyper-Taylorized labor process whose continually escalating productivity expectations heightens stress and exhaustion; a dormitory system that mixed workers from different production lines and home provinces in an attempt to forestall solidarity; and the absence of genuine collective representation, with rank-and-file workers distrustful of delegates of China's lone official trade union federation. This portrait is corroborated by other investigative profiles of mobile manufacturing at Foxconn and other facilities in China (Chan 2013; Chan et al. 2008; Nordbrand and de Haan 2009; Wilde and de Haan 2006; Litzinger 2013; Sandoval 2013).

Neither Foxconn nor China has a monopoly on harsh labor conditions in the sector. A national challenger for mobile component manufacture and final assembly is India, where Sriperumbudur, a 210-acre SEZ specialized in telecommunications, has been touted in national policy circles as a model worthy of "replication" (Cividep 2010, 9, 13). One critical report on this SEZ, which hosts Nokia and several of its suppliers, including Foxconn, gave special attention to precarious employment: temporary work, short-term contracts, recruitment agencies, apprenticeships and other insecurity-inducing features of nonstandard work arrangements are described as the "norm" in these mobile manufacturing workplaces (Cividep 2010, 6). A government task force charged with furthering India's telecommunications industry nominated greater employment flexibility as a policy objective, a trademark effort to attract foreign direct investment from communicative capitalists eager to attune labor supply to consumer demand (Cividep 2010, 13).

While that imperative makes a routine of skipping job to job, a qualitatively different labor of mobility familiar to the assembly workforce is migration. The boom in mobile phone manufacturing in China has as its precondition the transit of millions of workers from agricultural communities to low-wage export-processing conurbations (see Ngai 2004). When living in the cities its labor sustains, this proletarianized population is, however, excluded from social rights otherwise extended to longtime city dwellers. Internal migrants are governed by a national household registration system whose binary categories—rural or urban—were modulated to include a temporary metropolitan status so to allow SEZs like Shenzhen to meet their recruitment needs

Another facet of the "variegated citizenship" (Ong 2006) corresponding to the occupational hierarchy of high-tech is displayed in India. There too the contribution of migrants to the mobile industry has been integral. Of corporate hiring preferences, one Nokia SEZ employee offered: "They want only migrant workers. . . ." (in Cividep 2010, 27). Access to a "reserve army of cheap labour" has been identified as a pillar of mobile phone production in the Sriperumbudur area (2010, 22), with local job seekers marginalized in favor of those from far-flung areas who are deemed less likely to "voice their concerns and risk their sole source of livelihood" (2010, 32). Rather than the frictionless mobility conjured up by wireless advertising campaigns, it is the differential capacity for human movement that is leveraged by assembly capital for optimal labor control.

Even so, if the assumption was that a disproportionately young, temporary and migrant assembly workforce would be docile, then that has proven to be a dubious managerial calculation. Nokia, for example, was beset by three strikes in a single year at its factory in Sriperumbudur in 2009–10 (Fontanella-Khan 2010). Those conflicts revolved around wages and corporate retaliation, and concluded with the replacement of a company union with an independent one (Cividep 2010, 23–26). In China, autonomous labor unrest has surged in recent years (China Labour Bulletin 2014)—and has escalated at Foxconn itself (Chan, Pun, and Selden 2013, 108–111). Not only have mobile manufacturers such as Foxconn been pressured to raise wages (Foster 2010), but the circulation of labor struggles across sectors has been attributed in part to the use of mobile devices (Barboza and Bradsher 2010)—one of the subversive practices marking China's emerging "working-class network society" (Qiu 2009). As China Labour Bulletin (2014, 21) reports: "Workers . . . have proven themselves highly adept at using mobile social media tools to not just to (*sic*) organize protests but ensure that they come to the attention of the public, the traditional media and local government officials."

Although China remains a "gaping hole" in labor internationalism (Ross 2006, 263), the Foxconn workers in Guadalajara, Mexico, who held a vigil in solidarity with their Chinese counterparts offer an inkling of the transnationally coordinated labor action necessary to confront mobile capital (Chen 2010). Labor protest, coupled with planetary wage differentials, has, however, fed the restlessness of companies in the mobile sector: inland China and Vietnam are increasingly attractive lower-wage destinations (Chan et al. 2008, 20; Barboza 2010). Less well-known is that migrant work in the assembly of information technology now reaches into the European Union, the newest frontier of such production. In Pardubice and Kutna Hora, a hundred kilometers or so from Prague, a workforce of Bulgarians, Mongolians, Romanians, Poles, Ukrainians and Vietnamese assembles electronics for Foxconn (Andrijasevic and Sacchetto 2013). It is not only in geographical but also technological fixes that communicative capital is attempting to avoid its antagonists: Foxconn announced in 2011 its plan to add tens of thousands of robots to the shop floor (Branigan 2011); by 2013, it was reported that some twenty thousand were put to work in what is now characterized by Foxconn CEO Terry Gou as a "million robot army," installed in response to rising wages (Gou in Kan 2013). On this ever-shifting map of assembly labor at least one signal is clear: where capital goes so too does conflict.

DESIGN

Beyond the factory gates, the relationship between communicative capitalists and their connected global value subjects is increasingly mediated by mobile applications. The design of smartphone apps has been the most celebrated form of work along the circuit of wireless production. Reports that Apple's App Store had dispatched royalties tallying more than $6 billion (Streitfeld 2012) doubtless feeds the self-perception among some app writers that they are "gold diggers in the Klondike of the online world" (Bergvall-Kåreborn et al. n.d.). App development has been depicted as a dream job, which, with a little training and a lot of passion, can lead to "six-figure salaries and a bright future" (Tutelian 2011). Closer inspection, however, has revealed a more ambivalent picture of app employment.

Gorged by eager streams of venture capital, promising fledgling studios are morphing into familiar hierarchical structures of corporate high-tech. Mobile application development is "no longer the domain of IT nerds knocking up games in their bedrooms," but an "industry worth billions of pounds and employing hundreds of thousands of people worldwide" (King 2011). Fortunes of individual app developers are disproportionately shaped by the whims of companies creating operating systems for handsets. While this market was once dominated by manufacturers such as Nokia, Motorola, Samsung, Siemens and Sony Ericsson, first Apple, and then Google, brought about a different production model wherein application development was crowdsourced—and risk outsourced—to a global pool of workers, the corporate titans ensconcing themselves as gatekeepers of the app portal, enabling them to win whether an app is downloaded one million times or once.

According to preliminary research on app developers' working lives, design labor is characterized by a broad spectrum of contractual relationships, running from full-time, permanent work to a variety of flexible forms of employment, but it nonetheless tends toward "increasingly precarious employment and insecure prospects" (Bergvall-Kåreborn and Howcroft 2013, 965; see also Dyer-Witheford, forthcoming). Freelance or self-employed developers take on more risk, adopting entrepreneurial attitudes by necessity and bearing the brunt of failure as they navigate the inherent risks of the market. As one independent developer said, "[i]n the long run, working on your own products is more profitable than working on other people's products—but only if yours become a success" (Hollemans 2011). Indeed, app development has a lopsided income distribution profile: thousands of apps are launched every week, yet, notes the *New York Times*, "only a small minority of developers actually make a living by creating their own apps. . . ." (Streitfeld 2012). Linked to this IP lottery economy are the long-hours culture and the work/life blurring reported by high-end workers elsewhere in the tech sector (Bergvall-Kåreborn and Howcroft 2011, 2013).

At the extreme end of such flexible employment, online labor brokers such as oDesk and Elance (Caraway 2010) dole out small fragments of app development work. "Any job you can possibly think of can now be outsourced online," Freelancer.com CEO Matt Barrie boasted, adding that "[f]or the first time, small businesses can outsource projects [for] as little as

$30, and have delivered roughly the equivalent of $300 of western world [labor]" (Barrie in Mello 2010). Communicative capitalism has become deft in its ability to flexibly tap the world's digital workforces fuelling the breakneck development of the wireless frontier.

App developers are not all entrepreneurial, nor are mobile applications condemned to reinforce vested interests (de Peuter et al. 2014). Precarious freelance app developers formed the Android Developers Union and the App Developers Union to advocate for improved terms; these initiatives appear to have been short-lived, however. Where the autonomy of design labor has been more powerfully expressed is an emerging class of dissonant apps: GoodGuide allows consumers to scan barcodes to obtain environmental information on products; the U.S. Department of Labor's Timesheet app enables workers to track their own hours in response to mounting disputes revolving around unpaid back wages and inaccurately kept employer records (Hananel 2011); the British Trades Union Congress developed a Rights for Interns app, which allows interns to assess their internship for compliance with UK labor law; and hacktivist interventions such as artist Ricardo Dominguez's controversial Transborder Immigrant Tool was envisaged to provide tactical location-specific information to migrants seeking safe passage from Mexico to the U.S. (Dominguez 2011). Beyond the dedicated activist-oriented apps emerging is the now legendary subversive use of mobile communication—SMS, photo sharing, social networks—to catalyze, coordinate and circulate social conflict, from wage struggles among Chinese workers, to the occupations of the Arab Spring, through to Occupy Wall Street—wildly different outbreaks, to be sure, but all indicate that the genie of bottom-up, horizontal, social-movement connectivity has been let out of the bottle (Gerbaudo 2012).

MOBILE WORK

Approaching the labors of mobility from a materialist perspective requires bringing into view not only the labor that produces the smartphone, but also the transformation *of* labor through its hybrid connection to this device. Underscoring the importance of this relationship, the media theorist Franco Berardi (Bifo) suggests that the mobile phone acts as a kind of next-generation assembly line in the dispersed production process of an economy that increasingly trades in knowledge, cultural forms, and other forms of intellectual property (Berardi, 2012). Rendering labor infinitely accessible and hence always potentially available for work, the mobile phone is a vital communicative supplement to post-Fordist just-in-time employment. Entirely normalized, for example, is the roaming labor control that allows the flexworker—whatever its location—to be promptly called to a shift, and now workers bewildered by multiple jobs or erratic schedules can download an app like Shift Worker to get a grip on the flux.

More than a means to summon a subject to a workplace or to organize an unstable work itinerary, however, the mobile phone has increasingly
become a site for the performance of work as such. Distracted professionals hunched over smartphones at dinner on account of excessive workload,
managerial command, or misplaced affection are the cliché example of the
spatio-temporal extension of the workday via mobile telephony. But it is
experiments in so-called crowdsourcing that most strikingly indicate the
cybertarian possibilities emerging in the realm of mobile work.

By 2012, suggested the *Economist* ("Mobile Marvels" 2009, 13), there
would be close to two billion people in possession of a cell phone "but no
bank account." Pay-as-you-go feature phone users in the Global South constitute a swelling consumer segment. They also form a teeming labor pool, or
so proposes Jana (formerly txteagle), a Boston company founded in 2009 by
an academic entrepreneur at MIT. Jana dispatches simple info-tasks via text
message to what it perceives as an "untapped work force in developing countries" (Marwaha 2009). In a pilot project, Jana recruited bilingual Kenyans
to use their mobiles to receive and translate words to regional dialect for a
multinational telco localizing a handset interface (Bennett 2010). The company, which is present in more than fifty countries, has more recently distributed market research surveys on behalf of international brands seeking
an informed incursion into otherwise opaque "frontier economies" (Eagle
cited in Jidenma 2011). Helping to steer the market's possible expansion
path into the periphery, these txt-workers are, in a recursive loop, com-
pensated in airtime, whereby access to communication itself functions as a
measure of value exchanged for quantum doses of immaterial labor. While
Jana's scheme might appear farfetched, the start-up received a $15 million
venture capital injection in 2013 to further its effort to leverage mobile users
as mobile workers (Alspach 2013).

Digital piecework platforms distributing bite-size jobs to Web workers
for micropayment proliferated in the first decade of the 2000s. An early
entrant, Amazon.com's Mechanical Turk, allowed remote workers to complete small chores, like label a product's color for an online catalog, for
pennies. LivePerson brokered in expertise, with programmers, psychologists
and other professionals paid by the minute for online chat. iStockphoto
tapped amateur digital photography to fill an Internet archive of discount
images. These and kindred "cloud labor" markets soon offered mobile
applications (Leonard 2013). Other companies catered to mobile-specific
tasks, such as the smartphone users hired out by uTest or Mob4Hire as app
testers. Launched in 2010, Fancy Hands promises to "automate all the boring parts of your life" (in Leonard 2013) by outsourcing domestic chores to
a combined workforce of Internet temps which is estimated to stand at over
six million and to be at least doubling in size every year (Crowdsourcing.
org in Lionbridge 2013).

The micro-work system that has perhaps best demonstrated the possibility of harnessing smartphone portability to feed content to the mobile

Internet is, however, Gigwalk. The California company boasts to have retooled the iPhone as a vehicle for earning a "second paycheck" (Gigwalk 2014) by coining "mobile temp work" (Kim 2011). Using GPS to match a "job" to a nearby smartphone, Gigwalk allocates location-based data-gathering assignments, such as snapping pictures of menus for MenuPages.com, of businesses for Bing map results, and of the coordinates of Manhattan billboards for an ad agency database (Hann 2011; Kelly 2011). Motorola deployed Gigwalk for reconnaissance on its products' placement in third-party retail stores. As technological intermediary between communicative capital and mobile labor, Gigwalk profits from itinerant temps usually earning a few dollars per job via PayPal—a bargain, the app's developers estimate, compared to the costs conventionally associated with procuring local mapping data (Alsever 2011; Hann 2011). Piloted in a handful of major U.S. cities, in 2011 Gigwalk claimed its registered mobile workers had "travelled the equivalent of five trips to the moon and back" (Kim 2011). Since then, its workforce has grown fourfold, to over 200,000 "users" (Gorsht 2014). Rather than direct the movement of the dispersed smartphone population its software aggregates, Gigwalk encourages subscribers to merely "work gigs into their regular routines" (Alsever 2011). The implication that Gigwalk is nonintrusive is, paradoxically, only sensible at a point when metropolitan subjectivity is already so comprehensively enveloped by accumulation and commercial surveillance as to render it thoroughly banal.

Gigwalk and Jana provide a preliminary glimpse of how mobile devices might facilitate a continually refreshed visual cartography of consumer culture. But the phenomenon of mobile users doubling as pieceworkers marks a mutation in labor-capital relations as well. "The new phenomenon," as Berardi (2005) has argued, "is not the precarious character of the job market, but the technical and cultural conditions in which info-labour is made precarious." If the technical conditions include the mobile communication and reticular workflow software which provide a material infrastructure for the flexible deployment of labor, then the cultural conditions encompass increasingly diffuse generic competencies in basic computer use, Web navigation, and cellular interface. These techno-cultural conditions are multiplying the options available to capital for accessing immaterial labor power—while absolving itself of costly, and always potentially disruptive, employees. With "info-work," Berardi speculated, "there is no longer a need to have bought over a person for eight hours a day indefinitely." Mobile crowdsourcing perfectly bears out that "[c]apital no longer recruits people, but buys packets of time, separated from their interchangeable and occasional bearers" (Berardi 2005).

But if one tendency is for mobile work to be fractalized, another is for it to be unbounded, a dynamic most clearly displayed in the domain of unpaid mobile work. The smartphone is opening new portals for the continual reproduction of Web 2.0 and the shaping of purchasing flows within the broader economy by users engaging in what Tiziana Terranova (2004) called "free labour." Here the unpaid activity that produces and reproduces the Web is harvested by emerging titans of communicative capitalism. Google, Facebook, Apple and other contenders are vying for supremacy in the development of mobile applications allowing for instant status updates, geo-locational advertising, and the construction of "walled gardens" for the assembly, striation, and sale of what Dallas Smythe (1981) called the "audience commodity." The user-driven review site Yelp signals this trend, aggregating the helpful tips, photographic documentation and accounts of gastronomic and shopping adventures uploaded by a portion of the site's 120 million-plus monthly visitors (Yelp 2014).

Rough indications of the potential value of this user-generated activity comes in the form of projected advertising revenues from social media, forecast to surpass $10 billion (US) by 2017 (BI Intelligence in *Business Insider*: "Why Social Media . . ." 2013). The market for services which "locate the customer in space" (Goggin 2006, 197) and triangulate this with user-generated content and individual consumer proclivities is estimated to grow from slightly over US$2 billion in 2009 to US$14 billion in 2014 (Zaledon in Manzerolle 2010). The "mobile audience commodity" (Manzerolle 2010) is a serious prize, with the acquisitions of mobile advertising firms Quattro Wireless by Apple ($275 million) and AdMob by Google ($750 million) pointing to the value of such end-user attention and activity as smartphone density increases ("Phoney War" 2010).

SUPPORT

A fifth moment in the circuit of exploitation is support. Once a smartphone is in the customer's hands, the telecommunications companies that charge for access to the mobile Internet offer accessibility, responsiveness and personalized attention, but such promises are expensive to deliver. Their solution, the same as that adopted in a range of industries since the 1990s, is the call center, and as a result a strikingly large and highly disciplined workforce toils in panoptic settings in order to enable the 4G "user experience." Along with the financial sector, telecommunications firms are the largest call center employers in the world (Holman et al. 2007, 7). Working at what Huws (2009) described as the "interface" between communicative capitalism and the high-value customer segment accessing the Web through its smartphones, call center workers address billing complaints, resolve technology failures, sell services and collect on overdue accounts.

As one of the fastest-growing jobs in the information society, such work is highly feminized, affective and precarious, offering a vivid cross-section of the composition of immaterial labor today. In the telecommunications industry, call center workers tend to occupy the bottom rung in the corporate hierarchy and make the lowest pay. Like some of the other workers animating the circuit of wireless production, their employment conditions vary widely depending on whether they work "in-house" or for an outsourcing company. The former positions are more likely to be unionized, better paid and more secure, while the latter are more disciplinary in their organization, pay close to minimum wage, and see constant and endemic high levels of employee turnover, or churn (Holman et al. 2007). Not surprisingly, this dividing line is an important generator of labor unrest.

Relations between employers and workers are full of friction in the telecommunications sector, and call center workers have been at the heart of some of the more serious conflicts (Brophy 2010). Deregulation of the telecommunications market has dealt a serious blow to organized labor in recent decades, and since wireless is the fastest-growing portion of this market, companies have acted decisively to keep their incumbent unions from organising workers in the sector (Shniad 2007; Doellgast 2012). Within this struggle, the outsourcing of call center work is common and acts as an important management stick in the face of labor resistance, contractual demands and union drives.

The terrain for these conflicts, like the circuit examined in this chapter, is global, with companies seeking to exploit state borders as they play workforces in different countries off against each other. In one example, Canadian company Telus outsourced call center work to the Philippines during its 2004 lockout of Telecommunication Workers Union of Canada employees (Mosco and McKercher 2008, 145). Deutsche Telecom, which accommodates its unions at home in Germany, has fiercely resisted organising drives by the Communication Workers of America in the United States at its T-Mobile call centers (Gaus 2011). When global wireless company Vodafone was the target of violent criticism (and looting) during the Egyptian uprising for cutting off wireless communication at the behest of the regime, the company routed its local call center work to faraway New Zealand. The transfer of customer support functions away from the home corporation and into the highly precarious outsourced sector happens within domestic borders as well however. In 2005 Egyptian telecom entrepreneur (and former Mubarak regime insider) Naguib Sawiris bought the wireless provider Wind in Italy, and he soon outsourced its call center operations to local company Omnia, which promptly put workers on less secure contracts.

These examples suggest that call centers offer capital the perfect technological assemblage through which to achieve mobility at the flick of a switch. The growth of call center support work has, however, been paralleled by the emergence of labor resistance in a variety of forms, some of it surprising in its ferocity. Workers at the very bottom of the telecom pyramid have engaged in everything from traditional organizing drives by established unions to subterranean acts of digital sabotage, crafting small-scale innovative tactics such as flexible strikes and developing transnational organizing campaigns. Snapshots of this unrest include the ongoing series of actions engaged in by employees at the above-mentioned Omnia call center, including strikes, demonstrations and even the reported kidnapping of the company's managing director in Milan, who was forced to answer workers' questions regarding late wages and temporary contracts (ANSA 2009). A self-organized collective of precariously employed workers in Rome answering outsourced calls from Telecom Italia Mobile repeatedly brought Europe's largest call center, Atesia, to its knees between 2005 and 2007 and forced changes in Italian labor law around flexible employment (Mattoni 2012). Since 2008, the CWA has partnered with ver.di, the telecom union on the German-owned company's home soil, forming the T-Mobile Workers Union, and in New Zealand, when Vodafone expanded its local operations in order to flee Egyptian unrest, it ran straight into the rank-and-file union Unite, which has organized workers at all three of its Auckland locations. Barriers to communicative capitalism's own mobility should not be exaggerated.

DISASSEMBLY

As Moore's Law extends its dominion over the technologies of mobile communication, the "unstated corollary" of its expansive logic is becoming increasingly difficult to ignore: our state-of-the-art machines are always on the verge of obsolescence (Carroll 2008). Communicative capitalism produces almost unthinkable technological refuse, and electronic waste is the fastest-growing toxic stream on the planet today (Basel Action Network 2012). Adding up the various sources of electronic waste, the United Nations Environment Programme suggests, could total as much as fifty million tons of refuse a year (in Grossman 2010, 15). Sales of handsets and tablet computers were estimated at 1.8 billion in 2013 (Gartner in Lomas 2014) but these technologies get old quick. When the machines through which we access the mobile Internet are disposed of they bloat the international market for e-waste disposal, adding to the mountains of discarded gadgets leeching their poisonous externalities into air, soil and groundwater. On the wrong side of this "negative abundance" (Verzola 2009) are the laboring populations forced to make a living upon, and amid, this detritus.

Attention to the international flows of e-waste reveals the "previously disconnected geographies" of IT production, consumption and disposal (Vallauri 2009). Asia is the heart of electronics assembly today, yet the e-waste phenomenon spotlights the circularity and interconnectedness of global capitalism. After tidy profits on electronic junk is made by unscrupulous international e-waste brokers, at the "end" of the cycle of wireless accumulation (most notoriously in China, India and Ghana, but increasingly in other locales across Asia, Africa, Latin America and Eastern Europe) nightmarish scenarios are unfolding in which old and young have little choice but to engage in the noxious labor of e-waste disassembly, recuperating valuable metals from discarded technology by methods including burning with fire or dipping into vats of acid.

The global e-waste trade highlights the interconnections between the formal and informal economies supporting communicative capitalism. China's black-market e-waste industry is estimated to be worth $3.75 billion (US), importing around eight million tons of discarded electronics to the country each year (Nuwer 2014). In Guiyu, southern China, the emblematic site of global e-waste processing, the water is undrinkable, lead poisoning is endemic, high levels of such endocrine-disrupting chemicals have been found in children, and workers in recycling workshops suffer from skin, respiratory and gastrointestinal ailments (Grossman 2010, 17). This digital divide snakes through the Global North as well: U.S. government contractor Unicor uses inmate labor for electronics recycling in seven federal prisons. A United States Department of Justice (2010) report described prison staff and inmates working coated in toxic dust, and trailing heavy metals back to their homes and cellblocks at the end of their shift.

The circularity of global e-waste flows is arresting. Thanks to explosive demand, metals recycled from discarded electronics find their way back into the Chinese manufacturing process, from which they begin their return overseas to haunt—and sometimes kill—Western consumers. Poisonous leaded solder of the kind used in the manufacture of electronic circuit boards has popped up in American dollar-store jewelry and children's toys. As one of the American chemists responsible for tracing the poisons points out, "the U.S. right now is shipping large quantities of leaded materials to China, and China is the world's major manufacturing center . . . It's not all that surprising things are coming full circle and now we're getting contaminated products back" (cited in Carroll 2008). In this way the cycle of wireless accumulation runs its course, only to be renewed at the point of design upgrade, setting it in motion once again.

Yet if surveying the "toxic reciprocity" (Cutillo 2010) of commodity flows back and forth across the digital divide might well provoke despair, it is important to remember that the current visibility of e-waste as a serious problem and the scramble on the part of electronics manufacturers to develop strategies to deal with growing public criticism is the result of decades of grassroots struggle (Smith 2009). Reaching out along the very same circuits traversed by communicative capitalism's commodities, labor and environmental activists have linked up to hold manufacturers to account. The Silicon Valley Toxics Coalition, which initially emerged out of concerns over labor conditions in the Silicon Valley electronics manufacturing sector, has become part of a global network of organizations vowing to make communicative capitalism deal with its garbage, including, among others, the Electronics TakeBack Coalition in the U.S., PHASE II in Scotland, Asia Monitor Resource Centre in Hong Kong, TAVOI in Taiwan and CEREAL in Mexico (Smith 2009).

These groups are demanding extended producer responsibility, a policy which would hold manufacturers accountable for the full costs of their products at every stage in their lifecycle (a watered-down version of which has already been adopted in the European Union). But the ultimate goal of such groups is more ambitious. In 1999, just before the global justice movement erupted onto the scene, the Trans-Atlantic Network for Clean Production proposed a policy match to Moore's Law: "Each new generation of technical improvements in electronic products should include parallel and proportional improvements in environmental, health and safety, as well as social justice attributes" (in Smith 2009). Seemingly incompatible with the priorities of communicative capitalism, the NGO call for "fair and green mobile phones" delivered to network operators (SOMO 2010) requires connection to emerging discussions envisaging radically different political-economic orders.

MOBILIZING IMAGINARIES

The smartphone is a technical centerpiece of the imaginaries, ideologies, infrastructures and industries characterizing communicative capitalism. Necessary to this system's expansion is universalizing access to telecommunication networks, the achievement of which is touted as one of the great "mobile marvels" ("Mobile Marvels" 2009). What is left out in neoliberal narratives promoting the democratizing couplet of free markets and technological diffusion, however, are the radically unequal terms on which "global value subjects" have been integrated into the wireless frontier (Dyer-Witheford 2003).

The most prominent feature of the labors of mobility outlined in this chapter is their sheer heterogeneity: diggers use preindustrial techniques to extract ore in highly informal resource economies; proletarianized migrants perform hyper-Taylorized work in gargantuan factories where mobile devices are assembled; teams of entrepreneurial engineers design apps that are entered into the intellectual property lottery; precariously employed call center workers deploy scripted linguistic labor to manage mobile consumers; cellular subscribers do digital piecework and feed their collective intelligence and social networks into the mobile Internet economy; and e-waste workers scavenge through the debris of perpetual upgrade to eke out a subsistence livelihood on the "planet of slums" (Davis 2006).

The smartphone cybertariat is, then, an assemblage of extraordinary multiplicity. The collective workforce populating the circuit of exploitation that constitutes the mobile commodity displays the paradigmatic features of the contemporary innovation economy outlined by Paolo Virno (1996, 18–19): "a sort of 'umbrella' under which is replicated the entire history of labor: islands of mass workers, enclaves of professionals, swollen numbers of the self-employed, and new forms of workplace discipline and individual control. The modes of production that over time emerged one after the other are now," Virno added, "represented synchronically, almost as if at a world's fair."

So although mobile technologies currently sit among the commanding heights of communicative capitalism, a labor perspective on the smartphone makes clear that this economic formation is not defined by a single form of labor or process of production but is instead an exceedingly hybrid configuration: interdependence, co-presence, and indeed "unrestrained multiplication" (Virno 2004, 105) of strikingly different modes of labor organization, are characteristic features of even the most apparently avant-garde commodities of communicative capitalism. Failure to acknowledge this "eclectic coexistence" (ibid.) is to recapitulate the worst of New Economy thinking's blind spots.

A broad conception of communicative labor such as that presumed in this chapter poses enormous political challenges. Will the differentiation marking the smartphone cybertariat limit its resistance to easily contained outbreaks along the mobile circuit of exploitation? Or, can this variegated protagonist develop a counter-connectivity, linking across the barriers presented by distance, borders, and radically different cultures and contexts? Does the smartphone offer any organizing potential for a labor movement that has had difficulty establishing a presence in sectors identified with communicative capitalism? The mobile device is certainly an "ideal commodity" (Lee 1992; Kline et al. 2003) of communicative capitalism, crystallizing its technological, economic, cultural, and ecological dynamics. Yet the ideal-commodity concept, while a useful diagnostic tool, lacks a political correlate. By way of conclusion we gesture at the outlines of a possible mobile labor campaign which might fall under the rubric of a transnational "commodity unionism."[1]

The promise of building a labor campaign around the smartphone is significant. As the outpouring of grief at the death of Steve Jobs in 2011 made clear, the mobile device is currently the unrivalled sentimental icon of communicative capitalism. As the latter is not only an economic but also an ideological formation, puncturing its neoliberal claims of democracy achieved through technology is a key goal for labor and its allies. Some of the premises for its achievement are in place. Despite claims of their cynical detachment, mobile consumers in the overdeveloped world have begun to express concern for the plight of mobile assembly workers in the South, as projects such as makeITfair suggest, and while labor-oriented economic justice campaigns led by privileged social groups from the North warrant the scrutiny they have received (see Erçel 2006), it would be foolhardy to dismiss a growing desire on the part of consumers to understand the production context (and thus confront the fetishism) of the commodities they are implicated in. As the brand sabotage pioneered by the global justice movement demonstrated so effectively, the more ubiquitous communicative capitalism's promises are, paradoxically, the more vulnerable these become to skillfully wrought counter-information campaigns. The battle is asymmetrical, but not impossible. This much is suggested by the case of Fairphone. Growing out of an NGO campaign around conflict minerals, the Netherlands-based Fairphone is a crowdfunded social enterprise that began small-batch producing its own "fair" smartphone in 2013. "The greater goal," as reported in *The Globe and Mail*, "is to change entire economic systems and in this way the phone acts as a research project into the globalized economy" (Pett 2014).

Consumer-oriented action, brand sabotage and social enterprise, while important features of a campaign to alter the smartphone's relations of production, cannot be set on the same plane as the self-organization of labor. The multitudinous workforce animating the cycle of wireless accumulation calls out for experimentation in the development of new organizational forms and connections between and among moments on the circuit. Along these lines one relevant model is that of the world company council, wherein union representatives from different countries associated with a single multinational (including workers employed by contractors) meet to share knowledge and strategize (see Ferus-Comelo 2008, 148). Global gatherings of this sort are a necessary starting point to confront a mobile industry engaged in planetary labor arbitrage. Transnationally coordinated labor actions could be strengthened through the support of aspiring international worker organizations (e.g. UNI Global Union) willing to take a cue from thus-far isolated demonstrations of global solidarity, such as the earlier mentioned vigil held by Mexican Foxconn workers for their Chinese counterparts.

Italian literary collective Wu Ming (2011) write:

> If we are "within and against" the network, perhaps we can find a way to forge alliances with those who are exploited upstream. A global alliance between "digital activists," cognitive workers and electronics manufacturing employees would be, for the owners of networks, the most fearful thing. The forms of this alliance, obviously, remain to be discovered. (author's translation)

One part of exploring the possibilities of such an alliance is mapping networked labors of mobility and, in particular, identifying hot points of conflict and potential bottlenecks in the supply chain. Ultimately, however, a mobile labor campaign must be clear that communicative capitalism's promises cannot be fulfilled within its own framework: assumptions about accessibility are compromised by the secrecy of working conditions in mobile manufacturing supply chains; the enhanced mobility promised to smartphone consumers has its condition of possibility in the restricted movement of migrant workers assembling mobiles; imaginaries of a smooth space of communication downplay the fractures and frictions within the labor force enabling the mobile Internet. Pointing out such contradictions is not difficult. If "social imaginaries" are "pragmatic templates for social *practice*" (Herman 2010, 190), then the greater challenge confronting the smartphone cybertariat is that of constructing oppositional social imaginaries, imaginaries that would help to convert the utopian impulse animating "whatever" connectivity (Dean 2010, 68–69) into political solidarity.

NOTE

1. Our use of the term *commodity unionism* is distinct from that of Annunziato (1990), who uses it to refer to the commodification of collective representation.

SOURCES

Alexander, R. "Which is the World's Biggest Employer?" *The Economist*. March 19, 2012. Accessed 6 March 2014. www.bbc.com/news/magazine-17429786

Alsever, J. "Gigwalk: Make Money on Your Morning Commute." CNNMoney. June 22, 2011. Accessed March 6, 2014. http://money.cnn.com/2011/06/22/technology/gigwalk/

Alspach, K. "Mobile Startup Jana Raises $15M from Communications Giant Publicis." Boston Business Journal. July 19, 2013. Accessed March 3, 2014. www.biz journals.com/boston/blog/startups/2013/07/jana-raises-15m-from-publicis.html

Andrijasevic, R., and D. Sacchetto. "China May be Far Away but Foxconn is on our Doorstep." openDemocracy. June 5, 2013. Accessed March 6, 2014. www.opendemocracy.net/rutvica-andrijasevic-devi-sacchetto/china-may-be-far-away-but-foxconn-is-on-our-doorstep

Annunziato, F. R. "Commodity Unionism." *Rethinking Marxism*. no. 3 (1990): 8–33.

ANSA. "Recession: Omnia Network Late with Wages, Protest in Turin." April 3, 2009. Available through Factiva database. Accessed August 21, 2014. www.factiva.com/

Apple. *Supplier Responsibility: 2014 Progress Report*. January 2014. Accessed March 7, 2014. http://images.apple.com/supplier-responsibility/pdf/Apple_SR_2014_Progress_Report.pdf

Barboza, D. "Supply Chain for iPhone Highlights Costs in China." *The New York Times*. July 5, 2010. Accessed March 6, 2014. www.nytimes.com/2010/07/06/technology/06iphone.html

Barboza, D., and K. Bradsher. "In China, Labor Movement Enabled by Technology." *The New York Time*. June 16, 2010. Accessed March 6, 2014. www.nytimes.com/2010/06/17/business/global/17strike.html

Basel Action Network. "2012 Annual Report." 2012. Accessed March 3, 2014. www.ban.org/files/BAN_2012_Annual_Report.pdf

Bennett, D. "The End of the Office . . . , and the Future of Work." *The Boston Globe*. January 17, 2010. Accessed March 6, 2014. www.boston.com/bostonglobe/ideas/articles/2010/01/17/the_end_of_the_office_an d_the_future_of_work/

Berardi, F. "Info-Labour and Precarisation." Generation Online. 2005. Accessed March 6, 2014. www.generation-online.org/t/tinfolabour.html

Berardi, F. *The Uprising: Poetry and Finance*. Cambridge, MA: MIT Press, 2012.

Bergvall-Kåreborn, B., and D. Howcroft. "Crowdsourcing and ICT Work: A Study of Apple and Google Mobile Phone Application Developers." 2011. Accessed March 6, 2014. http://pure.ltu.se/portal/files/33142288/Abstract_till_FALF_2011_klar.pdf

Bergvall-Kåreborn, B., and D. Howcroft. "'The Future's Bright, the Future's Mobile': A Study of Apple and Google Mobile Application Developers." *Work, Employment and Society*. no. 6 (2013): 964–981.

Bergvall-Kåreborn, B., M. Björn, D. Chincholle, C. Hägglund, S. Larsson, and M. Nyberg. "The Pioneers of Mobile Application Development: Profiles of Android and Iphone Developers." n.d. B. SATIN Project. Accessed March 6, 2014. www.satinproject.eu/sites/default/files/Thepioneersofmobileapplications development.pdf

Branigan, T. "Taiwan iPhone Manufacturer Replaces Chinese Workers With Robots." *The Guardian.* August 1, 2011. Accessed March 6, 2014. www.theguardian.com/world/2011/aug/01/foxconn-robots-replace-chinese-workers

Brophy, E. "The Subterranean Stream: Communicative Capitalism and Call Centre Labour." *Ephemera.* no. 4 (2010): 470–483. Accessed March 6, 2014. www.ephemerajournal.org/contribution/subterranean-stream-communicative-capitalism-and-call-centre-labour

Caraway, B. "Online Labour Markets: An Inquiry into oDesk Providers." *Work Organisation, Labour & Globalisation.* 4, no. 2 (2010): 111–125.

Carroll, C. "High Tech Trash: Will Your Discarded TV End Up in a Ditch in Ghana?." *National Geographic,* January 2008. Accessed March 6, 2014. http://ngm.nationalgeographic.com/2008/01/high-tech-trash/carroll-text

Castells, M. *Communication Power.* New York: Oxford University Press, 2009.

Chan, J. "A Suicide Survivor: The Life of a Chinese Worker." *New Technology, Work and Employment.* no. 2 (2013): 84–99.

Chan, J., N. Pun and M. Selden. "The Politics of Global Production: Apple, Foxconn, and China's New Working Class." *New Technology, Work and Employment.* no. 2 (2013): 100–115.

Chan, J., E. Haan, S. Nordbrand and A. Torstensson. *Silenced to Deliver: Mobile Phone Manufacturing in China and the Philippines.* Amsterdam: SOMO and SwedWatch, 2008.

Chang, G. "Is Foxconn Fleeing China?" *Forbes.* February 24, 2013. Accessed March 6, 2014. www.forbes.com/sites/gordonchang/2013/02/24/is-foxconn-fleeing-china-sure-looks-like-it/

Chen, M. "Foxconn's Global Empire Reflects a New Breed of Sweatshop." *These Times.* October 13, 2010. Accessed March 6, 2014. http://inthesetimes.com/working/entry/6550/foxconns_global_empire_reflects_a_new_br eed_of_sweatshop

China Labour Bulletin. "Searching for the Union: The Workers' Movement in China, 2011–13." 2014. Accessed March 6, 2014. www.clb.org.hk/en/sites/default/files/Image/research_report/Searching for the Union.pdf

Cividep. *Changing Industrial Relations in India's Mobile Phone Manufacturing Industry.* Amsterdam: SOMO, 2010.

Custer, C. "China Made 1.2 Billion Mobile Phones Last Year." *Tech in Asia.* February 7, 2013. Accessed March 6, 2014. www.techinasia.com/china-12-billion-mobile-phones-year/

Cutillo, C. "A Toxic Reciprocity: Examining the E-Waste Stream from the United States to China." *Harvard Law and Policy Review.* 2010. Accessed October 5, 2014. http://hlpronline.com

Davis, M. *Planet of Slums.* London: Verso, 2006.

Dean, J. *Democracy and Other Neoliberal Fantasies.* Durham: Duke University Press, 2009.

Dean, J. *Blog Theory: Feedback and Capture in the Circuits of Drive.* Cambridge, UK: Polity, 2010.

de Peuter, G., E. Brophy, and N. S. Cohen. "Locating Labour in Mobile Media Studies." In *The Routledge Companion to Mobile Media*, edited by G. Goggin and L. Hjorth, 439–449. London: Routledge, 2014.

Doellgast, V. *Disintegrating Democracy at Work: Labor Unions and the Future of Good Jobs in the Service Economy.* Ithaca: Cornell University Press, 2012.

Dominguez, R. "Transgressing Virtual Geographies: Maryam Monalisa Gharavi Interviews Ricardo Dominguez." *Darkmatter,* 2011. Accessed March 6, 2014. www.darkmatter101.org/site/2011/07/27/transgressing-virtual-geographies/

Doya, D. M. "Congo's Mining Ban Fails to Break Links With Armed Groups," Institute Says." *Bloomberg News.* March 1, 2011. Accessed March 6, 2014.

www.bloomberg.com/news/2011-03–01/congo-s-mining-ban-fails-to-break-links-with-armed-groups-institute-says.html

Dyer-Witheford, N. "The New Combinations: Revolt of the Global Value Subjects." *The New Centennial Review.* no. 3 (2003): 155–200.

Dyer-Witheford, N. "App Worker." In *The Imaginary App*, edited by P. D. Miller and S. Matviyenko. Cambridge, MA: MIT Press (forthcoming).

El Akkad, O., and I. Marlow. "Commodity Boom: The Smartphone's Global Price War." *The Globe and Mail.* June 1, 2013, B4.

Engels, F. "The Condition of the Working Class in England," 1845. Accessed March 6, 2014. https://www.marxists.org/archive/marx//works/1845/condition-working-class/

Erçel, K. "Orientalization of Exploitation: A Class-Analytical Critique of the Sweatshop Discourse."*Rethinking Marxism.* no. 2 (2006): 289–306.

Ferus-Comelo, A. "Mission Impossible? Raising Labor Standards in the ICT Sector." *Labour Studies Journal.* no. 2 (2008): 141–162.

Fontanella-Khan, J. "Pay Strike Hits Nokia Handset Hub in India." *Financial Times.* July 14, 2010. Accessed March 6, 2014. www.ft.com/intl/cms/s/0/365dbaae-8f6f-11df-ac5d-00144feab49a.html

Foster, P. "China Faces Wave of Strikes After Foxconn Pay Rise." *The Telegraph.* June 10, 2010. Accessed March 6, 2014. www.telegraph.co.uk/finance/china-business/7818406/China-faces-wave-of-strikes-after-Foxconn-pay-rise.html

Gaus, M. "AT&T Bid for T-Mobile Boosts Union Drive."*Labour Notes.* May 3, 2011. Accessed March 6, 2014. www.labornotes.org/2011/05/att-bid-t-mobile-boosts-union-drive

Gerbaudo, P. *Tweets and the Streets: Social Media and Contemporary Activism.* London: Pluto, 2012.

Gigwalk. "Contact Us." Accessed March 6, 2014. http://gigwalk.com/contactus

Global Witness. "'The Hill Belongs to Them': The Need for International Action on Congo's Conflict Minerals Trade." 2010. Accessed March 6, 2014. www.global witness.org/sites/default/files/library/The hill belongs to%2 0them141210.pdf

Goggin, G. *Cell Phone Culture: Mobile Technology in Everyday Life.* New York: Routledge, 2006.

Gorsht, R. "Can Unemployment Be Eradicated By Rethinking Jobs?" *Forbes.* February 2, 2014. Accessed March 3, 2014. www.forbes.com/sites/sap/2014/02/02/can-unemployment-be-eradicated-by-rethinking-jobs/

Grossman, E. "Tackling High-Tech Trash: The E-Waste Explosion and What We Can Do About It." *Demos.* 2010. Accessed March 6, 2014. www.demos.org/sites/default/files/publications/High_Tech_Trash-Demos.pdf

Hananel, S. "Smartphone App Lets Workers Track Wages." *Associated Press.* May 22, 2011. Accessed August 21, 2014. www.today.com/id/43128963/ns/today-today_tech/t/smartphone-app-lets-workers-track-wages/

Hann, C. "Finding Customers Ahead of a Startup Launch." *Entrepreneur.* July 26, 2011. Accessed March 6, 2014. www.entrepreneur.com/article/220018

Harvey, D. *The Enigma of Capital and the Crises of Capitalism.* Oxford University Press, 2010.

Herman, A. "'The Network We All Dream Of': Manifest Dreams of Connectivity and Communication or, Social Imaginaries of the Wireless Commons." In *The Wireless Spectrum: The Politics, Practices, and Poetics of Mobile Media*, edited by B. Crow, M. Longford and K. Sawchuck, 187–198. Toronto: University of Toronto Press, 2010.

Hogg, J. "DR Congo Says to Lift Eastern Mining Ban March 10." *Reuters.* March 2, 2011. Accessed March 6, 2014. www.reuters.com/article/2011/03/02/ozabs-congo-democratic-mining-idAFJOE72101X20110302

Hollemans, M. "Life as a Freelance iPhone Developer." 2011. Accessed March 6, 2014. www.hollance.com/2011/02/life-as-a-freelance-iphone-developer/

Holman, D., R. Batt and U. Holtgrewe. "The Global Call Centre Report: International Perspectives on Management and Employment." 2007. Accessed March 6, 2014. www.ilr.cornell.edu/globalcallcenter/upload/GCC-Intl-REpt-US-Version.pdf

Huws, U. *The Making of a Cybertariat: Virtual Work in a Real World.* New York: Monthly Review Press, 2003.

Huws, U. "Working at the Interface: Call Centre Labour in a Global Economy." *Work Organisation, Labour and Globalisation.* no. 1 (2009): 1–8.

IDC. "Worldwide Mobile Phone Market Maintains its Growth Trajectory." *Reuters.* February 1, 2012. Accessed March 6, 2014. www.reuters.com/article/2012/02/01/idUS263098 01-Feb-2012 BW20120201

International Alert. "The Role of the Exploitation of Natural Resources in Fuelling and Prolonging Crises in the Eastern DRC." 2010. Accessed March 6, 2014. www.international-alert.org/sites/default/files/publications/Natural_Resources_Jan_10.pdf

Jidenma, N. "Mobile Startup Txteagle Uses SMS to Gather Consumer Insights." *The Next Web.* April 25, 2011. Accessed March 6, 2014. http://thenextweb.com/africa/2011/04/25/mobile-startup-txteagle-uses-sms-to-gather-consumer-insights-in-emerging-markets/

Johnson, J. "My Gadget Guilt." *Wired.* March 2011, 96–103.

Kan, M. "Foxconn to Speed Up 'Robot Army' Deployment." *PC World.* June 26, 2013. Accessed March 20, 2014. www.pcworld.com/article/2043026/foxconn-to-speed-up-robot-army-deployment-20000-robots-already-in-its-factories.html

Kelly, M. "GigWalk adds Microsoft, Offers 110K Paying Odd Jobs." *Mobile Beat.* July 19, 2011. http://venturebeat.com/2011/07/19/gigwalk-microsoft-gigs/ (accessed March 6, 2014).

Kim, R. "Gigwalk Finds a Market for Mobile Temp Work; Signs Microsoft as a Client." *GigaOM.* 2011. Accessed March 6, 2014. http://gigaom.com/2011/07/19/gigwalk/

King, M. "A Working Life: The App Developer." *The Guardian.* March 5, 2011. Accessed March 6, 2014. www.theguardian.com/money/2011/mar/05/working-life-app-developer

Kline, S., N. Dyer-Witheford and G. de Peuter. *Digital Play: The Interaction of Culture, Technology, and Marketing.* Montreal and Kingston: McGill-Queen's University Press, 2003.

Lee, M. J. *Consumer Culture Reborn: The Cultural Politics of Consumption.* London: Routledge, 1992.

Leonard, A. "The Internet's Destroying Work—And Turning the Old Middle Class into the New Proletariat." *Salon.* July 12, 2013. Accessed March 6, 2014. www.salon.com/2013/07/12/the_new_proletariat_workers_of_the_cloud

Lim, W., and T. Culpan. "Hon Hai, Foxconn International Tumble After Earnings Miss Estimates." *Bloomberg News.* August 31, 2010. Accessed March 6, 2014. www.bloomberg.com/news/2010-08-31/hon-hai-foxconn-international-shares-tumble-after-earnings-miss-estimates.html

Lionbridge. "The Crowd in the Cloud: Exploring the Future of Outsourcing." *Massolution.* January 2013. Accessed March 6, 2014. www.lionbridge.com/files/2012/11/Lionbridge-White-Paper_The-Crowd-in-the-Cloud-final.pdf

Litzinger, R. "The Labor Question in China: Apple and Beyond." *South Atlantic Quarterly.* no. 1 (2013): 172–178.

Lomas, N. "Gartner: Smartphone Sales Finally Beat Out Dumb Phone Sales Globally In 2013, With 968M Units Sold." *TechCrunch.* February 13, 2014.

Accessed March 3, 2014. http://techcrunch.com/2014/02/13/smartphones-outsell-dumb-phones-globally

Ma, T. "China and Congo's Coltan Connection." *Project 2049 Institute.* 2010. Accessed March 6, 2014. http://project2049.net/documents/china_and_congos_coltan_connection.pdf

Mantz, J. W. "Improvisational Economies: Coltan Production in the Eastern Congo." *Social Anthropology/Anthropologie Sociale.* no. 1 (2008): 34–50.

Manzerolle, V. "Mobilizing the Audience Commodity: Digital Labour in a Wireless World." *Ephemera.* no. 3–4 (2010): 455–469.

Marlow, I., and O. El Akkad. "Smartphones: Blood Stains at Our Fingertips." *The Globe and Mail.* December 3, 2010. Accessed March 6, 2014. www.theglobeandmail.com/technology/smartphones-blood-stains-at-our-fingertips/article1318713/

Marwaha, A. "New Service is All in a Day's SMS." *BBC World Service.* February 11, 2009. Accessed March 6, 2014. http://news.bbc.co.uk/2/hi/technology/7881931.stm

Marx, K. *Grundrisse.* New York: Penguin, 1857.

Mattoni, A. *Media Practices and Protest Politics: How Precarious Workers Mobilize.* London: Ashgate, 2012.

McKercher, C., and V. Mosco. "Editorial: The Labouring of Communication." *Canadian Journal of Communication.* no. 3 (2006): 493–497.

Mello, J. P. "Tech Wars Fuel Freelance Jobs." *PC World.* July 6, 2010. Accessed March 6, 2014. www.pcworld.com/article/200537/Tech_Wars_Fuel_Freelance_Jobs.html

"Mobile Marvels." *The Economist.* September 26, 2009. Accessed March 6, 2014. www.economist.com/node/14483896

Mosco, V., and C. McKercher. *The Laboring of Communication: Will Knowledge Workers of the World Unite?* Lanham: Lexington Books, 2008.

Ngai, P. "Women Workers and Precarious Employment in Shenzhen Special Economic Zone, China." *Gender and Development.* no. 2 (2004): 29–36.

Nordbrand, S., and E. de Haan. *Mobile Phone Production in China: A Follow-up Report on Two Suppliers in Guangdong.* Amsterdam: SOMO and SwedWatch, 2009.

Nuwer, R. "Eight Million Tons of Illegal E-Waste Is Smuggled Into China Each Year." *Smithsonian.com.* February 28, 2014. Accessed August 21, 2014. www.smithsonianmag.com/smart-news/eight-million-tons-illegal-e-waste-smuggled-china-each-year-180949930/

Ong, A. *Neoliberalism as Exception: Mutations in Citizenship and Sovereignty.* Durham: Duke University Press, 2006.

Pett, S. "Meet Fairphone: A Phone Company Turning Protest into a Disruptive Product." *The Globe and Mail.* February 14, 2014. Accessed March 6, 2014. www.theglobeandmail.com/technology/tech-news/meet-fairphone-a-phone-company-turning-protest-into-a-disruptive-product/article16901664/

"Phoney War: Google Unveils a Rival to the iPhone." *The Economist.* 2010. Accessed March 6, 2014. www.economist.com/node/15210100

Pole Institute. *The Coltan Phenomenon: How a Rare Mineral has Changed the Life of the Population of War-Torn North Kivu Province in the East of the Democratic Republic of Congo.* Goma: Pole Institute, 2002.

Pöyhönen, P., K. Bjurling and J. Cuvelier. "Voices from the Inside: Local Views on Mining Reform in Eastern DR Congo." *Finnwatch and Swedwatch.* 2010. Accessed March 6, 2014. http://somo.nl/publications-en/Publication_3586/at_download/fullfile

Qiu, Jack Linchuan. *Working-Class Network Society.* Cambridge, MA: MIT Press, 2009.

Ross, A. *Fast Boat to China: High-Tech Outsourcing and the Consequences of Free Trade: Lessons from Shanghai.* New York: Vintage, 2006.

Sandoval, M. "Foxconned Labour as the Dark Side of the Information Age: Working Conditions at Apple's Contract Manufacturers in China." *tripleC.* no. 2 (2013): 318–347.

Schiller, D. *How to Think About Information.* Urbana: University of Illinois Press, 2007.

Shniad, S. "Neo-liberalism and Its Impact in the Telecommunications Industry: One Trade Unionist's Perspective." *Knowledge Workers in the Information Society,* edited by C. McKercher and V. Mosco, 299–310. Lanham: Lexington Books, 2007.

Smith, J., and J. Mantz. "Do Cellular Phones Dream of Civil War? The Mystification of Production and the Consequences of Technology Fetishism in the Eastern Congo." *Inclusion and Exclusion in the Global Arena,* edited by M. Kirsch, 71–93. New York: Routledge, 2006.

Smith, T. "Why We are 'Challenging the Chip': The Challenges of Sustainability in Electronics." *International Review of Information Ethics.* no. 11 (2009): 9–15.

Smythe, D. *Dependency Road: Communications, Capitalism, Consciousness, and Canada.* Norwood: Adlex, 1981.

SOMO. "Thousands of Consumers Across Europe Demand Fair & Green Mobile Phones." *Make IT Fair.* December 6, 2010. Accessed March 6, 2014. http://makeitfair.org/en/the-facts/news/thousands-of-consumers-across-europe-demand-fair-green-mobile-phones

Streitfeld, D. "As Boom Lures App Creators, Tough Part is Making a Living." *The New York Times.* November 17, 2012. Accessed January 9, 2013. www.nytimes.com/2012/11/18/business/as-boom-lures-app-creators-tough-part-is-making-a-living.html

Students and Scholars Against Corporate Misbehaviour.*Workers as Machines: Military Management in Foxconn.* Hong Kong: SACOM, 2010.

Terranova, T. *Network Culture: Politics for the Digital Age.* London: Pluto, 2004.

Tutelian, L. "Dream Jobs: Six Figure Salaries and a Bright Future." *Yahoo Finance.* March 3, 2011. Accessed March 6, 2014. http://finance.yahoo.com/news/pf_article_112224.html

United Nations Development Programme. "Table 1: Human Development Index and its Components." 2013. Accessed August 21, 2014. https://data.undp.org/dataset/Table-1-Human-Development-Index-and-its-components/wxub-qc5k

United States Department of Justice Oversight and Review Division. "A Review of Federal Prison Industries Electronic-Waste Recycling Program." 2010. Accessed March 6, 2014. www.justice.gov/oig/reports/BOP/o1010.pdf

Vallauri, U. "Beyond EWaste: Kenyan Creativity and Alternative Narratives in the Dialectic of End-of-Life." *International Review of Information Ethics.* no. 11 (2009): 20–24.

Verzola, R. "21st Century Political Economies: Beyond Information Abundance." *International Review of Information Ethics.* no. 11 (2009): 52–61.

Virno, P. *The Ambivalence of Disenchantment. Radical Thought in Italy: A Potential Politics,* trans. M. Turits. Minneapolis: University of Minnesota Press, 1996.

Virno, P. *A Grammar of the Multitude,* trans. I. Bertoletti, J. Cascaito, and A. Casson. New York: Semiotext(e), 2004.

Wagner, W. "iPad Factory in the Firing Line: Worker Suicides Have Electronics Maker Uneasy in China." *Spiegel Online.* May 28, 2010. Accessed March 6, 2014. www.spiegel.de/international/business/ipad-factory-in-the-firing-line-worker-suicides-have-electronics-maker-uneasy-in-china-a-697296.html

Wall, M. "2013: The Year We All Went 'Mobile.'" *BBC News.* December 19, 2013. Accessed March 5, 2014. www.bbc.com/news/business-25445906

"Why Social Media Advertising Is Set to Explode." *Business Insider.* June 6, 2013. Accessed March 3, 2014. www.businessinsider.com/social-media-advertising-set-to-explode-2013–55

Wilde, J., and E. de Haan. *The High Cost of Calling: Critical Issues in the Mobile Phone Industry.* Amsterdam and Stockholm: SOMO and SwedWatch, 2006.

Wu Ming. "Feticismo della Merce Digitale e Sfruttamento Nascosto: I Casi Amazon e Apple." 2011. Accessed March 6, 2014. www.wumingfoundation.com/giap/?p=5241

Yelp. "About Us." 2014. Accessed March 3, 2014. www.yelp.ca/about

York, G. "The Bleak Calculus of Congo's War Without End." *The Globe and Mail.* March 26, 2010. Accessed March 6, 2014. www.theglobeandmail.com/news/world/the-bleak-calculus-of-congos-war-without-end/article4318421/

Part II
Mobile Pasts and Futures

5 Wireless Pasts and Wired Futures

Ghislain Thibault

> Ether, having once failed as a concept is being reinvented. Information is the ultimate mediational ether. Light doesn't travel through space: it is information that travels through information . . . at a heavy price.
>
> —Sol Yurick, *Behold Metatron*

In 1914, when Thomas Edison was asked by the *New York Times* what the greatest invention had been since the electric light bulb, he scribbled "wireless" at the very top of a list published by the newspaper (Marshall 1914, 1–2). It was becoming usual in both vernacular and technical discourse to employ the term "wireless" as a noun, a short for "wireless telegraphy." And indeed, while electric telegraphy had been one the great marvels of the nineteenth century, a token of modernity, its evolution into a *wireless* system baffled many and fueled wild promises about its future applications.

It might be said that from this point two histories of wireless part. The first history is well-known and vast: it is one that traces the evolution of the dominant communication technologies of the twentieth century into wireless devices: radio, television, telephone and recently computing networks. However, when the discovery of 'wireless' was announced in the last decade of the nineteenth century, its future was by no means a linear one and it certainly was not limited to communications. The "wireless age" was a site of social, technical and cultural speculations about the new ethereal method of transmission. A counter-history to the long series of successful material technology, explored here, focuses on the social imaginaries around the potentialities of wireless (Taylor 2004). This chapter looks for the "imaginary" or "discursive" media (Huhtmamo 1997; Kluitenberg 2011) that never materialized into a working prototype or a commercialized technology. In particular, I explore some of the imagined scenarios concerning the possibility of wireless power following the discovery of the Hertzian waves. Thus, this is a history of social and scientific desires, predictions and fantasies, a history of what Canales call "desired machines" (2011). Through an analysis of imagined models of wireless power transmission, from its first manifestation in the late nineteenth century to the renewed interest in its potential applications today, I hope to shed light on some of the tensions between materiality and mobility that mark our technological culture.

THE WAR ON WIRES

The commercial success of a family of new electrical technologies in the second half of the nineteenth century resulted in a proliferation of wires in urban and domestic life. The electric telegraph, the telephone, the electric trolley and the electric light were all wired technologies, and as they disseminated, so did intricate networks of wires of copper, steel, aluminum, nickel, iron. The electrification of industrial machinery further consolidated a network of production and distribution of electrical power and in the wake of the rapid and extensive electrification of the Western world, more wires appeared. In the urban context, wires were a symbol of the new modern connectivity. The new scale of social and economic relationships was made fully visible by the presence of the wires. Historians of technology have ably documented how the telegraph system was often explained using the metaphor of the human nervous system (Nye 1990; Morus 2000). Wires fostered continuous couplings and embodied the very infrastructure for a reticular social body. Countless observers resorted to this analogy to describe the interconnection between distant markets, such as this Living Age reporter who in 1845 viewed the magnetic telegraph as "a net-work of nerves of iron wire, strung with lightning, will ramify from the brain, New York, to the distant limbs and members: . . . Pittsburgh, Cincinnati, Louisville. . . ." (*The Magnetic Telegraph: Some of Its Results* 1845, 194). As material evidence of a rising networked society, wires were visible symbols of technological progress. The households of early telephone subscribers and electric light adopters were easily identifiable. In rural areas, too, the advent of poles and wires in a small town meant that regional phone companies deemed the place worthy of connection with the modern world (Fischer 1992). As a result, a dictionary of phrases and fables published in 1896 suggested completing Lucretius's three-age classification of human civilization (the stone, bronze and iron ages) with a fourth: the wire age, with its "telegraphs, by means of which well-nigh the whole earth is in intercommunication" (Brewer 1896, 22).

Cohabitation with wired technologies in social space was experienced with ambivalence. On the one hand, they made possible new kinds of connectivity that pushed the boundaries of time and space. As James Carey (1989) noted in his seminal work on the telegraph, the separation of transportation and communication was a main consequence of its invention, making modern communication appear ever more disembodied and immaterial. In reality though, the infrastructure that supported these seemingly ephemeral global interconnections was all too material and the plethora of wires threatened mobility in both urban and domestic spaces. Overhead wires and power lines overshadowed the streets of large cities in America. By the 1880s, wires had proliferated to such degree that they became a source of anxiety. The *New York Times* printed countless dramatic stories of deadly encounters with the wires: cases of buildings unexpectedly burning down, trees catching on fire and plenty of electrocutions were reported. Even firemen were "obstructed, frightened, and killed by the unseen force." Reporters noted that the seemingly harmless telegraph wires were "rendered just about as dangerous as the light wires" when they came in contact with their electrified counterpart, as happened in the case of a boy who was electrocuted after stepping on a fallen telegraph line on Pearl Street in 1890. Overhead wires added "new terrors to living" in New York and citizens became captives of their homes, unable to "venture upon their own housetops where those deadly wires may be." There was a generally held sentiment that streets and houses were disfigured by the unsightly wires. In fact, wires were so unpopular that, in 1883, editors of the newspaper insisted: "The telegraph and other wires must come down. The forests of masts and net-work [*sic*] of iron wires must disappear."[1]

The perception of wires as dangerously imperfect conduits for the invisible force of electricity did serve to assert the legitimacy of a new group of experts: electrical engineers and electricians (Marvin 1988). One advertisement from the United Edison Manufacturing Company in the early 1890s warned that buildings were "better not wired than wired poorly." Only expert linemen were permitted to install, repair or touch wires (and this was also true of telephone, telegraph and electrical lines), keeping amateurs at bay. The risks associated with the rapid diffusion of modern electric media extended even to those who might have chosen to not be 'wired.' Since major companies refused to bury wires underground due to higher installation and maintenance costs, the number of wires and poles in cities grew quickly and soon impeded people's ability to circulate freely in public spaces. Ironically, increased connectivity was quickly coming to threaten social mobility. Some city councils negotiated with utilities companies to allow the use of sidewalk space to install poles in exchange for free electricity or telephone service (Fischer 1992, 133–5). In New York, a citizen's right to "free passage," as the *New York Times* described it, was a growing concern. A Bureau of Incumbrances had been created in 1871 in order to solve the problem of street and sidewalk obstruction caused by empty carts, fruits stalls, paving stones, boxes and other rubbish. Poles and wires came under the Bureau's mandate following the Great Blizzard of March 1888, when a dangerous mess of downed wires left in its wake was deemed a public nuisance. Popular discontent with the overhead wires climaxed. An annual report to the U.S. Secretary of War noted how the blizzard "clearly shows the necessity of placing all wires underground," and urged the Capitol to bury them as soon as possible in order to set a good example for electrical companies. In 1899, New York Mayor Hugh Grant transformed his political campaign into a "pole and wires crusade" and tasked the Bureau of Incumbrances with putting wires underground as soon as he came into office. That year alone, the bureau reported burying 12,308 miles of wires operated and owned by various phone, telegraph and electric light companies.[2]

By 1890, the public's initial fascination with and excitement about the scientific innovation that allowed disembodied communications, artificial lightning and rapid transportation was eclipsed by a growing awareness of the dangers posed by the material infrastructure these inventions required. In particular, the fragility, cost, unpredictability and potential hazards posed by the wires constituted significant disincentives to further expansion of the networks of distribution of both energy and information. In the context of this popular resentment against the materiality of wires, the idea of "wireless" came to be viewed a very practical solution to the problem of preserving the advantages of the new modern connectivity without impinging on the freedom of movement and action.

The lexicon of "wireless" was employed generously: even putting wires out of sight was sufficient to be deemed "wireless." The wire department of the city of Boston, for instance, claimed in 1897 to have "more wireless streets" as poles were cleared in various parts of the city. The wires themselves still existed, but their concealment acted to safeguard both mobility and connectivity in the eyes of the authorities.

The longing for a more radical solution was shortly answered. By the mid 1890s, electrical engineers claimed the discovery of a method for the transmission of messages without the recourse to wires, following experiments on the electromagnetic spectrum by several scientists and electricians (including Heinrich Hertz, William Crooke, Oliver Lodge, Nikola Tesla and Guglielmo Marconi). When it announced wireless telegraphy in 1897, the *New York Times* noted how the overabundance of scientific discoveries, inventions and consumer products that had seen the light of day in the last decades of the nineteenth century had made everyone "supercilious with regards to the novelties in science," but that this "languor may be stirred up at the prospect of telegraphing through air and wood and stone without so much as a copper wire to carry the message." Wireless telegraphy was clearly an invention in a class apart: as Susan Douglas (1987) noted, it was "miraculous" to many and it "bridged the chasm between science and metaphysics" (23). In contrast to Boston's wire department "wireless" achievement, the unwiring announced by electrical engineers and scientists was radical: wireless through the Hertzian waves was meant as wire*less*. Not only was it a practical solution to a real problem, it was also the realization of one of the fondest dreams of scientists at this time: the domestication of the ether.

ETHEREALIZATION AND PROGRESS

In his lengthy *A Study of History* published in 1934, British historian Arnold Toynbee attempted to identify the criteria indicating a civilization's growth. Approaching the question of the evolution of technology in history, Toynbee suggested that progress was tangled with a "law of progressive simplification": the shift from coal to oil fuel, from water to steam power, from steam to electric power, and from electric to wireless telegraph (Toynbee 1934). Toynbee notes that the simplification of a technical apparatus generates an improvement in efficiency and a reorganization of the social system that welcomes it. But Toynbee was not satisfied with the term "simplification" to designate this law:

> "Simplification" is a negative word. It connotes omission and elimination; whereas, in the concrete examples of the phenomenon from which we have inferred the validity of the law, the ultimate effect which the law produces by its operation is not a diminution, but an enhancement of practical efficiency or of aesthetic satisfaction or of intellectual understanding or of godlike love. In fact, the result is not a loss but a gain.
>
> (Toynbee 1934, 135)

The neologism "etherealization," he thought, was "more illuminating" (Toynbee 1934 135).[3] Etherealization can be understood as a movement from a material plane, or state, toward an ethereal one. The concept suggests that "dematerialization" should not be the opposing principle of "materialization." On the contrary, a transfer of social, aesthetic and spiritual forces accompany the transformation or reduction of matter. Thus true progress may not be found in machines being smaller and simpler, but in the subjective illumination that the new form fosters.

Wireless is perhaps the illustration *par excellence* of etherealization, and it responds to the meaning of the concept in various ways. First, the transition from wired telegraphy to wireless telegraphy, accomplished in the early twentieth century, was *literally* understood as a transition from the world of matter to that of the ether, this subtle and imponderable medium that had been a source of speculation among scientists and philosophers for centuries (Galison 2003). At the time the telegraph loses its wires, there is no doubt in the mind of the many observers, starting with engineers at the Marconi company, that the medium in which electromagnetic signals are sent is the mythical ether (Hong 2001, 59). Some naturally designated the new invention as "ethereal telegraphy" in the late 1890s, and although physicists abandoned luminiferous ether theories at the turn of the twentieth century (in part because of Einstein's famous claim in 1905 that it was superfluous to his principle of relativity), the term *ether* subsisted in the vernacular to describe the medium in which radio waves are broadcasted. Discursively and technically, then, the simplification of the technology of telegraphy had undergone a literal process of etherealization.

For the nonscientists, wireless telegraphy, a system simplified by the disappearance of annoying wires, was surely a testimony of the progress of science and technology; but the proximity of the new invention with the ether tapped into a much more profound aspiration, that of the harnessing of a new order of physical reality. While the webs of wires of telegraph and telephone exchanges could be explained with an analogy that had an equivalent in the natural world—the human nervous system—the etherealized telegraphy challenged any existing models about the living world. As a result, the disembodied communications of the wireless telegraph, along with new techniques of visual perception like photography and radiography, fueled spiritualist claims about telepathy and telekinesis (Davis 1999; Peters 1999; Stolow 2013), perhaps the closest analogies to the domain of ethereal.

The introduction of wireless telegraphy thus cannot be reduced to the mundane satisfaction that wires had, at last, disappeared. The significance of the etherealization of telegraphy in fact stimulated an heterogeneous range of "social imaginaries" (Taylor 2004). It instilled the imagining of what this new, phenomenologically imperceptible order of reality could offer for science and technology, and for everyday life as well. It reorganized both the science and the industry of electrical engineering, and opened up the commercial practices that would become the broadcasting industry. It also, and perhaps most importantly, finished to unsettle preexisting conceptions about physical reality and uplifted metaphysical claims in the popular and scientific discourse. Wireless became a space of speculation and potentialities—a space of 'virtuality.' And it was in that space that imaginaries about wireless power transmission took form.

THE DREAMS OF BROADCASTING POWER

The engineer Nikola Tesla (1856–1943) was one of the greatest promoters of the idea of wireless power transmission in the late nineteenth century. His invention of the polyphase system in the late 1880s was an impetus for the electrification of America by providing a technical means to transmit electricity over longer distances than existing systems could at the time. When the Edison Electric Lights Company provided electricity for street lightning in New York City in the 1880s, the company had to install dynamos right in the heart of Manhattan. Indeed, the lower tension of direct current distribution forced electric companies to build multiple stations in the densely populated areas where electricity was utilized the most. Such stations, as one British engineer noted in 1890, meant "the annoyance of machinery placed in a place where it is undesirable" for the public (Hopkinson 1890). By comparison, the alternating current system allowed Westinghouse engineers in 1895 to send power to the city of Buffalo from Niagara Falls, located more than twenty miles away (Hughes 1983, 135–139). The adoption of the alternating current system as a standard in the community of electrical experts transformed the model of electricity distribution: the sites of production of electrical power could be spatially distanced from the sites of consumption. As Hirsch and Sovacool recently noted, the history of electric utility system has been punctuated by a long process of making "its product largely invisible, both in its manufacture and physical manifestation. To many people, electricity remains an unseen and unthought-of commodity" (2013, 706). If this is generally true for most aspects of the technological system, poles and wires were (and are) still the exception to this regime of invisibility.

In 1891, at the height of the crusade against wires and before the announcement of wireless telegraphy, Tesla announced to the press that he was working toward perfecting a system of electric lighting that could be powered without wires. He was proposing to do with electrical power what would soon be done to telegraphy: getting rid of wires. A reporter told readers (with some pragmatism) that the system, known as the "Tesla Glow," would "appeal to the layman" because "the walls and ceilings will not be defaced by wires."[4] As international attention shifted to focus on Marconi's experiments in wireless telegraphy in 1897, Tesla multiplied his claims about a double-barreled system of wireless telegraphy *and* wireless power. In a patent filed in September 1897 on the transmission of electrical energy without wires, he described how wireless power transmission may be achieved: a coil would elevate the voltage inside a vacuum to such a level that a spark gap exciting an inductor would produce bursts of high-frequency currents of such "character and magnitude as to cause thereby a current to traverse elevated strata of the air between the point of generation and a distant point at which the energy is to be received and utilized" (1897). In contrast with Marconi, who limited his experiments to the transmission of pulse signals in the electromagnetic spectrum, Tesla envisaged that the same "medium" (Tesla refused to use the term ether[5]) was also suited for long distances, high tension, energy transfer.

Wireless power was Tesla's "fondest dream" (1897). He famously experimented with wireless transmission and high-frequency currents in Colorado at the end of 1899 (Carlson 2013). By 1900, he sought financial support to build his "World System," the construction of which was initiated on Long Island, New York, before being halted by the lack of capital. Tesla's project was further undermined by Marconi's first transatlantic radio transmission in 1901. This success, in turn, consolidated the fate of wireless with information transmission rather than power transmission. By 1902, Tesla was unable to pay the employees working on the Long Island tower, and he was late in paying Colorado Springs electricity bills from 1899. Carpenters deserted the site on July 19.[6]

As Marconi was asserting priority in the invention of wireless telegraphy, Tesla was battling to have his own system of wireless transmission of energy recognized as such. If there seems to be no historical evidence that he succeeded in creating a wireless power apparatus, Tesla recurrently declared that his system was near completion. These announcements were usually made to popular press reporters, who in turn got accustomed with the personage. A *Time* magazine article from 1931 recounts how Tesla even near the end of his life was still "rehashing" the "same old subject—Broadcasted

Power." Another article commented that the "wizard's dream" had still not come true when he turned eighty.[7]

In the early twentieth century, wireless power was not just the fantasy of a lone electrician whose fame was passing. While deeply associated with Tesla in the minds of many, predictions, experimental attempts and claims of priority about the invention were rather frequent. Consider a few examples. In 1901, news came from London that engineers had unveil the "secret" of wireless power system, "a new thing and a great thing." In May 1907, Sir Hugh Bell, who was ironically president of the Iron and Steel Institute, predicted wireless power would soon be a reality. That same year a high school student from Worchester, Massachusetts, claimed to be able to send power wirelessly over a distance of nine hundred feet. In April 1908, electrical engineer Dr. Frederick H. Millener claimed he had powered a truck from a distance. A wirelessly powered boat was reportedly seen on the Danube in Vienna in December 1911, and on February 10, 1914, a special cable announced that Marconi himself was able to light electric lamps from a distance of six miles in London.[8] These predictions, announcements and claims of invention of wireless power do not exclusively reflect the swiftness of the press when it came to publicizing scientific discovery despite proper experimental demonstration. On the contrary, in many cases, these announcements were made without a fanfare, as if wireless power was yet another footnote to the marvels of modernity. The detachment of the press and the claims themselves show that wireless power was almost already a commonplace idea. It had turned into a plausible expectation of the advancement of science and this expectation carried with it its own "normative notions and images" (Taylor 2004, 24).

The various descriptions of an imagined wireless power shared common features. I focus here on three: centralization, control and standardization. First, a shared view of wireless power was that the passage from wired to wireless would increase the centralization of networks of distribution of electricity. Tesla's World System, for instance, consisted of one large, conic tower mounted by a giant round magnifier that could transmit power wirelessly, regardless of distance. Tesla claimed on various occasions that only a few of these towers would suffice to provide large amounts of electricity throughout the globe.[9] Should that energy be freely distributed across the globe, it was, indeed, an economically flawed model. However, the model on which Tesla's wireless power relied was a highly centralized one: ownership of one of these towers could mean access to a global (and no longer local or national) market and control over both production and distribution. In theory, it was a very lucrative venture and many noted it as a commercial opportunity. "Want to make some money?" asked a reporter for the Chicago Tribune in 1912, "then invent wireless power" and "don't waste your time with wireless telegraphy or the wireless telephone . . . Wireless transmission of power is bigger and more important than either" (Diamond 1912). In 1925, a book entitled *What's Wanted* included wireless power as a needed invention from inventors (Price 1925). In 1927, the magazine *Popular Science Monthly* called "radio power," the "next great invention" (Armagnac 1927). The article was accompanied by a representation of transmitting towers depicted with the vocabulary of radio broadcasting: a central emitter in the center and a mass of receivers tapping from the periphery. Echoing the changes in audience that were taken place a time in the transition from telegraphy to radio—the rise of the concept of mass—reception of electricity in this proposed model evolved from being individual and local to being collective and global. In 1927, the chief of the Radio Laboratory at the Bureau of Standards attempted to set the record straight concerning wireless power transmission: "It may well be," he said, "that directive radio will never bring wireless power transmission." He added that while being feasible technically, "it would be the most inefficient thing in the world, and not even the wealth of Henry Ford would suffice to pay for the enormous transmitting station that would be required" (Dellinger 1927, F9). But even this prognostic about the impossibility of wireless power was portraying, with its *singular* station, a potentially highly centralized and monopolistic system.

The second feature shared by many of those who envisioned wireless power was greater control over machinery and engines powered by electricity. Like many of his contemporaries, Tesla believed that the possession of a superweapon would put an end to wars (Davis 1999, 88–89). Wirelessly powered warfare was just such weapon. In the years leading to and during World War I, he described how a wirelessly powered "manless airship" could "produce destructive effects at a distance."[10] Remotely controlled and unmanned warfare appealed to the military. In 1900, a United States Naval attaché claimed to have witnessed the workings of an invention by English engineer Cecil Varicas that was to be offered to his government, consisting of a method for "steering torpedoes and submarine craft by means of a wireless electrical device" using the Marconi system. A Navy commander noted in 1901 that the advantage of such a system was "the absolute control of the torpedoes at all times." With the possibility of controlling motion from a distance, crewless battleships were perceived to be a strategic military advantage. Some argued that warfare should "guided by the cool judgment and steady hand of a man safe ashore, instead of by the nervous hand of a man who knows he is going to almost certain death."[11] In 1910, the "aeronautic editor" of *Scientific American* pointed out that an "airship driven by wireless power will put an end to future wars," while suggesting that the newly created Carnegie Foundation for the Promotion of International Peace should back such a project.[12] Fifty years later, cybernetics engineers and scientists recognized the importance of open channels of communication for the self-regulation of machines (antiaircraft guns, radar and later drones). In the cybernetics model, the control is achieved through communication. The warfare engines dreamt of in the first half of the twentieth century were, in contrast, literal *perpetuum mobile* that turned machines into inexhaustible *automata*, self-regulated through information channels *and* self-supplied by energy transfer channels.

From a commercial perspective, control from a distance also had the potential to increase the existing distance between markets of energy production and markets of consumption, a tendency that Tesla's own alternating current system had initiated. Speculating on the realization of wireless power, a *Toronto Globe* reporter joyfully prognosticated in 1934 that "gasoline and copper wires industries . . . would have to take it 'on the chin'" ("Power Without Wires" 1934). In this model, the conditions of production were rendered more obscure, and certainly less visible to consumers.

A third major imagined feature of the etherealization of electric power was standardization. As the broadcasting model of radio took shape, the new neologisms of mass media were used to describe wireless power: expressions such as "radio power" or "broadcast power" appear in vernacular in the 1920s. This was not a mere analogy. In fact, the transmission of power would have relied, in the models of many, on the same medium as that of radio signals: the electromagnetic spectrum. The signals of early telegraphic systems were electrical impulses, and most developments in wireless telegraphy were made within the electrical engineering community. Information and energy in the late nineteenth century were part of the same ecology and their separation was artificial. Perhaps it was the etherealization of telegraphy that stabilized a view of the electromagnetic spectrum as a medium of *information* transmission, when the electromagnetic radiation occurring in the spectrum falls under energy phenomena. Tesla's view of wireless transmission systematically supported such an energetic view of the ether. His World System conflated all types of transmission into one wireless system: clock, positioning system, telephony, telegraphy, picture transmission, music and, of course, power. The system also hinted at a possible standardization of sound, text and image into one common, universal language.

WIRELESS POWER TODAY

Today, the enduring hype around everything "Wi-Fi" and "mobile" is a sign that wireless—or "wirelessness" as Adrian Mackenzie puts it (2010)— is more than a scientific sensation. It has become a quality, a feature of technology that is serving a pressing commercial and social necessity: mobility. Communication practices and social relations demand to be utterly mobile and wires—literally—get in the way. But to talk of *wireless* phones, computers and tablets is merely a figure of speech. Our devices may be mobile, but everyday experience reminds any user that using computers, cell phones and tablets requires a short halt to a wall socket, this relic of the industrial age.

After the long and steady etherealization of information technology— telegraphy, telephony, and computing—the industry tells us that now is the time for electricity to join in. In 2007, a team of MIT researchers succeeded in powering a light bulb without wire from a distance of seven feet. The invention was described by *Technological Review* in 2008 as one of the ten technologies most likely to change the way we live. In 2010, the Consumer Electronics Show in Las Vegas exhibited the first wirelessly powered plasma television. Wireless charging technologies have also burgeoned since. The industry is coming together to develop open standards for wireless charging around several organizations: the Wireless Power Consortium (2009), the Power Matters Alliance (2012) or the Alliance for Wireless Power (2012). Members in these industry organization include large consumers electronics and telecommunication providers (Samsung, AT&T, Sony), but also a few retailers who might entertain offering wireless charging on site (Starbucks Corporation, for instance).[13] In January 2014, the Massachusetts-based company WiTricity announced that they were ready to move on with the commercialization of their wireless charging system for smartphones and other consumer electronics.

Wireless power transmission is entering an already-cramped medium, the ether. Despite generally using the lower frequencies of the electromagnetic spectrum, and thus not impeding on the frequencies of telecommunications, wireless power remains an electromagnetic phenomenon. When Nikola Tesla and others were imagining their wireless power system, the ether was perceived as an infinite resource. Its imperceptibility and speculative nature had intensified a view of the ether as inexhaustible. The history of wireless telecommunications offers a more problematic conception of the true "ethereality" of wireless. As is well-known, the rhetoric behind radio regulations in the 1920s started articulating the ether as a finite resource (Douglas 1987; Hong 2001). Access to the ether was limited; frequencies were fragmented and sold. From an absolute, limitless and mythical substance filling all space, the ether was reconstructed as a finite medium. As Joe Milutis puts it, "the creation of real estate in the ether, through electromagnetic spectrum allocation and the proliferation of networks, is the most dramatic transvaluation that the world has recently undergone" (2003). The quantitative leap in data transmission in the spectrum, thanks to the wireless hype and generations of more demanding mobile technology, is so rapid and intense that electronic engineers are preparing for a potential saturation. This concern with the sustainability of the ether was evoked in the early days of wireless computing. In his "From the Ether" column published for *InfoWorld* magazine, Bob Metcalfe (who is also the engineer behind the Ethernet cable) was one of the first to make a sortie against the wireless trend, which he predicted would completely and permanently flop. He argued, "it is an ecologically unsound waste of energy to broadcast bits in all directions when they need to be received in only one. The ether is too scarce to be wasted on nonbroadcast communications, and it won't be" (1993b, 48). Nicolas Negroponte echoed this same argument in his famous book *Being Digital*:

as soon as we use the ether for higher-power telecommunications and broadcast, however, we have to be very careful that signals do not interfere with each other. We must be willing to live in predetermined parts of the spectrum, and we cannot use the ether piggishly. We must use it as efficiently as possible. Unlike fiber, we cannot manufacture any more of it. Nature did that once.

(Negroponte 1995, 24)

Both information technology experts saw the shift toward wireless as a process that would limit and constrain transmissions instead of deploying endless and ubiquitous connections. Network clogging, congestion, collapses, and e-waste management are issues the industry is starting to weight in on. Metcalfe wrote, "cutting all these cords and cables is exciting, but isn't inevitable . . . [E]ven if I'm wrong about the *permanent* shortage of real ether, wires will be keeping us civilized for a very long time" (1993a, 48). He concluded somewhere else: "Wire up your home and stay there" (1993b, 56). The growing traffic in the ether does not discourage wireless power developers from moving in. Some promoters argue that their technology does not collide with the electromagnetic spectrum, using low range resonance in magnetic fields that are "not very busy" (Schatz 2011), at least for now.

The future convergence of information and energy transmissions in the same medium reactivates some of the anxieties that echo the introduction of electric devices in the domestic and urban life during the industrial revolution. Only this time, it is the ethereality rather than the materiality of transmissions that ignites fears. Unwiring is no longer the solution for the potential harm found in wires; on the contrary, the invisibly of the networks of transmission is that which instil apprehension. The phenomenon of "electromagnetic hypersensitivity" is perhaps the best contemporary expression of this fear. Answering these growing concerns, the World Health Organization has been surveying the scientific literature on the effects of electromagnetic files on health since 1996 through the International Electromagnetic Fields Project. For many energy moving through the air brings up safety concerns and the industry prepares to answer those concerns. Once again, the lessons of Toynbee's etherealization are replayed here: as wired technologies lose their wires, they are not rendered "immaterial," simply the material structures transmutes into another form that open up new challenges. In this case, the passage from material structures (wallsockets, wires, poles) to pervasive, invisible and ubiquitous *infra*structures does not alleviate fears; rather, it amplifies them.

The promises of preserving and enhancing mobile culture seems key to the social acceptation of wireless power. In various advertisements promoting wireless technology over the Internet, access to a source of power is depicted as the major physical constraint to fully disembodied practices of communication. Currently the main application of wireless power, as promoted by many of the companies pioneering in the field, is the charging of consumer products (smartphones, televisions, tablets) and various household devices (wireless kitchen appliances, for instance). In turn, it answers the exigency of mobility that the information technology industry has carefully marketed in the last two decades. The industry of wireless power builds on mobility through a general catchphrase that cuts across most marketing discourses selling the innovation: wireless power finally cuts the last cord to the material world. In 2008, PowerMat's CEO Ran Poliakine argued "Recharging without the tangle and hassle of wires is the next logical step in the continuing evolution of wireless devices" (Poliakine 2011). Advertising material for PowerBeam, a U.S.-based company in wireless power, argues that technologies are not "truly" wireless until we are rid of all wires. Above all, say the company's advertisers, wireless power is "freedom from the wires" (PowerBeam Inc. 2011). Even the story that led to the experiments on wireless power at MIT in 2007 by Marin Soljačić, professor of physics and one of the founders of WiTricity, is told through the lenses of an ordinary mobile phone user, experiencing the limits of battery autonomy. Rather than a story of fundamental scientific research, it portrays the eureka movement when Soljačić, awakened at night for the umpteenth time by his cell phone battery needing recharging, resolved to pursue experiments in wireless charging (Schatz 2011). The imagined usages of wireless power by promoters appear to first serve the ideals of portability, autonomy and mobility of consumer electronics.

CONCLUSION

The historical analysis of wireless power imaginaries provided in this chapter may further serve to approach the tension between materiality and mobility in contemporary digital culture. First, the imagined systems of wireless power have participated to preserving the historical and cultural proximity between energy and information. As Linda Henderson and Bruce Clarke ably note, a "conceptual crossover from energy to information" took place at the turn of the twentieth century. The "heavy technologies of the earlier industrial period" and their "monumental energy gradually shifted toward the lighter structures of high technologies" (Clarke and Henderson 2002, 2). Those in the early twentieth century who speculated on the possibilities for energy production, distribution and consumption to move into the lighter structures offered an opposing voice to the prevalence of information. Tesla's stubborn persistence in referring to "wireless power" when he was challenging Marconi on the priority to the invention of radio is telling: it is as if he was not conscious of, or refused, the profound cultural transition into the "information age." As he and others were speculating on the capacity of the ether to be a common medium for both energetic and informational transfers, or using the lexicon of radio to describe how an apparatus of wireless power would work, they were keeping the frontier between information and energy fully open and fluid. As we have seen, the recent actualization of wireless power into a material technology might further challenge that boundary altogether, along with the growing awareness that the information economy does not exist outside "heavy" dimensions. Many digital media studies scholars, for instance, have been unpacking recently the deep materialities of the digital economy: e-waste, material conditions of production, labor issues.

The imaginaries of wireless power may also challenge the widely shared value of mobility. No longer a simple mantra for marketers of digital devices, mobility is now a social imperative, and wirelessness is still perceived today, as it was in the late nineteenth century, as the practical solution to the problem of the material aspects of technology infringing on the freedom of movement. The utter unwiring of telephones, tablets and computers with the advent of wireless charging is likely to be praised as the achievement of a new threshold in the liberation from the constraints of the material world. In fact, unwiring is not the only simplification (or to use Toynbee's term, etherealization) that information and communication technologies have undergone toward meeting the exigency of mobility. Their commercial success since the 1980s has relied on sustained processes of simplification: miniaturization (reduction of volume and weight, compression, batteries), digitization (standardization of language), naturalization ("user-friendly" interfaces, simulations of natural environments), delocalization (cloud computing, separation of production markets with consumer markets). Mobile communication, as a result, appears more disembodied, more immaterial than ever—an ambiguity captured by the misuse of the term "virtual" as the opposite of "real" (Lévy 2001). The seemingly dematerialized and distributed nature of digital networks has not, however, been exempt from politics of control and centralization (Galloway 2004). On the contrary, the history of radio broadcasting and wireless computing shows that the passage to the ether centralized communication. Tesla's imagined model of wireless power is helpful once again: while his discourse legitimating the invention suggested utopian dreams of open and distributive models of energy production and consumption, the material system of wireless power he imagined theoretically reinforced, on the contrary, centralization and control.

Wireless power was not perceived by the amateurs or experts envisioning it as a dematerialization; on the contrary, their models of what wireless power ought to be supported specific material arrangements, institutions, economic models, scientific theories, social norms, cultural ideals and personal dreams. As yet another passage from the wired world to the ether seems imminent, we might keep in mind for wireless what H. G. Wells famously said of railways: "Before every engine, as it were, trots the ghost of a superseded horse" (1902, 11).

NOTES

1. "Sudden Death in the Air" (1883, November 4), *New York Times*, p. 6; "The Deadly Electric Wires. The System of Stringing them Overhead Must Be Abolished" (1883, November 19), *New York Times*, p. 2; "The Monopoly of the Streets" (1883, January 31), *New York Times*, p. 4; "Dazed by Electricity" (1890, September 16), *New York Times*, p. 8.

2. "United Edison Manufacturing Company" (1892) [Advertisement], *The World Almanac*, New York: Press Publishing Company; United States Army Corps of Engineers (1889), Annual Report of the Chief of Engineers, United States Army, to the Secretary of War for the year 1889, Washington, DC: Government Printing Office; "The Pole and Wire Crusade" (1889, April 28), *New York Times*, p. 4.; New York Board of Electrical Control (1890, February 7), "Report of the New York Board of Electrical Control," *The Electrical Engineer. A Weekly Journal of Electrical Engineering*, 5: 116–118.

3. The reception of Toynbee's concept of etherealization was mixed. A contemporary reviewer finds the term "unfortunate" and concludes that with his "fluid psychoid concepts Toynbee has not progressed far beyond many another philosopher of history" (Kroeber 1943: 295). Others, like Marshall McLuhan, have found the concept somewhat useful. McLuhan recognized the similarity between Toynbee's etherealization and Buckminster-Fuller's "ephemerealization," and he will use the term to name the general principle behind the software revolution, and more generally to describe the capacity to "do more and more with less and less" (1973). Lewis Mumford utilized the expression in 1938 in *What Is City?* and later in *The Condition of Man* (1944), however without mention to Toynbee's work. Finally, Christian Kerslake (2008) analyzes how some concepts in Gilles Deleuze may have been informed by the philosopher's critical reading of Toynbee.

4. "Wireless Electric Lamps. Mr. Tesla's Experiments with High-Frequency Alternations" (1891, July 9), *New York Times*, n.p.

5. Despite the wide circulation of ether theories in physics and electrical science at the time, Tesla was careful to only rarely refer to the word "ether." He made use of "air," "atmosphere," "gaseous medium filling all space" or the "upper strata" in lieu of the ethereal vocabulary, which he considered "too vague." In reaction to Roentgen's discovery of X-rays in 1896, attributed to vibrations of the ether, Tesla wrote to the press that he preferred using "the term 'primary matter,' for, although the expression 'ether' conveys a perfectly definite idea to the scientific mind, there exists, nevertheless, much vagueness as to the structure of this medium." Nikola Tesla (Electrical Review, December 1, 1896), in Popović et al., *Articles, Selected Works*, p. 223.

6. Friction at Wardenclyffe. Tesla's Carpenters Quit Work on Saturday" (1902, July 14), *Brooklyn* Eagle, p. 9; "Colorado News Item" (1904, June 17), *Littleton Independent*, p. 8.

7. On Tesla asserting to the press that he was the inventor of radio, see for instance: N. Tesla (1907, October 25), "Letters to the Editor. Tesla on Wireless," *New York Tribune*, p. 7. See also "Tesla at 75" (1931, July 20), *Time Magazine,* pp. 27–30; "Wizard's Dream Is Unfulfilled in His 80th Year. Nikola Tesla, Inventor, Celebrates 79th Birthday—Faith Unbroken" (1936, October 13), *The Globe*, p. 3.

8. "Claim Secret of Wireless Power. English Electricians Declare They Have Discovered Sure Method" (1901, October 18), *Chicago Daily Tribune*, p. 4; "Wireless Power. The Most Important Scientific Forecast of the Week" (1907, May 19), *New York Tribune*, p. B3; "Sends Power by Wireless: Nineteen-Year-Old Boy Asserts He Has Worked Motors at 900 Feet" (1907, December 20), *New*

York Times, p. 4; "Wireless Power Boat" (1911, December 31), *Boston Daily Globe*, n.p; "Lights Lamps by Wireless" (1914, February 10), *The Washington Post*, p. 1; "Light and Power Via Radio Possible" (1927, June 26), *The Washington Post*, p. SM3; "Power Without Wires" (1934, April 26), *The Globe*, p. 1.

9. "Earth Will Speak to the Planets, Scientist Predicts" (1931, July 11), *The Globe*, p. 3.

10. Winans, Richard Maxwell (1912, March 3), "Wireless Power," *New York Tribune*, pp. 3–4, 16; "Tesla's New Device Like Bolts of Thor." (1915, December 8), *New York Times,* p. 8.

11. "Passing of the Torpedoes: Lieut. Commander Colwell Says New Wireless Electric Steering Device May Revolutionize Defenses" (1900, January 18), *New York Times*, p. 3 "New Method of Steering Torpedoes" (1900, March 11), *The Atlanta Constitution*, p. A3; "Steering Torpedoes by Wireless System: Successful Demonstration with Model of the Gardner Invention" (1901, December 11), *New York Times,* p. 8.

12. "How Can Carnegie Fund Be Best Used to Promote Peace?" (1910, December 25), *New York Tribune*, p. A.

13. Membership to both associations was retrieved in February 2014 from their respective websites. Wireless Power Consortium (2011), "Wireless Power Consortium," accessed on February 12, 2014, from www.wirelesspower consortium.com; and Power Matters Alliance (2013), "Our Members. Directory," accessed on February 16, 2014 from www.powermatters.org. Another organization, the Alliance for Wireless Power, merged with the Power Matters Alliance on February 11, 2014.

SOURCES

Armagnac, A. P. "Wireless Power: The Next Great Invention." *Popular Science Monthly.* no. 1 (1927).

Brewer, E. C. *Dictionary of Phrase and Fable: Giving the Derivation, Source, or Origin of Common Phrases, Allusions, and Words that Have a Tale to Tell.* Philadelphia: Henry Altemus Company, 1896.

Canales J. "Desired Machines: Cinema and the World in Its Own Image." *Science in Context.* no. 24 (2011): 329–359.

Carey, J. W. *Communication as Culture: Essays on Media and Society.* New York: Unwin Hyman, 1989.

Carlson, W. B. *Tesla: Inventor of the Electrical Age.* Princeton: Princeton University Press, 2013.

Clarke, B., and L. D. Henderson (eds.). *From Energy to Information: Representation in Science and Technology, Art, and Literature.* Stanford: Stanford University Press, 2002.

Davis, E. *TechGnosis: Myth, Magic and Mysticism in the Age of Information.* New York: Harmony Books, 1999.

Dellinger, J. H. "Trials in Directing Radio Waves Found Having Great Value." *The Washington Post.* April 2, 1927, F9.

Diamond, J. "Can You Invent Wireless Power?" *Chicago Tribune.* September 1, 1912, sec. F.

Douglas, S. J. *Inventing American Broadcasting, 1899–1922.* Baltimore: Johns Hopkins University Press, 1987.

"Etherial Telegraphy." *New York Times.* November 5, 1897.

Fischer, C. S. *America Calling: A Social History of the Telephone to 1940.* Berkeley: University of California Press., 1992.

Galison, P. *Einstein's Clocks, Poincaré's Maps: Empires of Time.* New York: W.W. Norton, 2003.

Galloway, A.R. *Protocol: How Control Exists after Decentralization.* Cambridge, MA: MIT Press, 2004.

Hirsh, R. H, and B.K. Sovacool. "Wind Turbines and Invisible Technology: Unarticulated Reasons for Local Opposition to Wind Energy." *Technology and Culture.* no. 4 (2013): 705–734.

Hong, S. *Wireless: From Marconi's Black-Box to the Audion.* Cambridge, MA: MIT Press, 2001.

Hopkinson, J. "English Authorities on High-Tension Underground Circuits, Discussion." *Electrical Engineer: A Weekly Review of Theoretical and Applied Electricity.* no. 9 (1890): 176–180.

Hughes, T.P. *Networks of Power: Electrification in Western Society, 1880–1930.* Baltimore: Johns Hopkins University Press, 1983.

Huhtamo, E. "From Kaleidoscomaniac to Cybernerd: Notes toward an Archaeology of the Media." *Leonardo.* no. 3 (1997): 221–224.

Kerslake, C. "Becoming against History: Deleuze, Toynbee and Vitalist Historiography." *Parrhesia.* no. 4 (2008): 17–48.

Kluitenberg, E. "On the Archaeology of Imaginary Media." In *Media Archaeology: Approaches, Applications, and Implications,* edited by E. Huhtamo and J. Parikka, 48–69. Berkeley: University of California Press, 2011.

Kroeber, A.L. "A Study of History. Arnold J. Toynbee." *American Anthropologist.* no. 2 (1943): 294–299.

Lévy, P. *Cyberculture* (vol. 4). Minneapolis: University of Minnesota Press, 2001.

Mackenzie, A. *Wirelessness: Radical Empiricism in Network Cultures.* Cambridge, MA: MIT Press, 2010.

Magnetic Telegraph, The: Some of Its Results. The Living Age. Boston: T.H. Carter and Company, 1845.

Marshall, E. "The Future Man Will Spend Less Time in Bed—Edison." *New York Times.* October 11, 1914.

Marvin, C. *When Old Technologies Were New: Thinking about Electric Communication in the Late Nineteenth Century.* New York: Oxford University Press, 1988.

Massumi, B. *Parables for the Virtual: Movement, Affect, Sensation.* Durham: Duke University Press, 2002.

McLuhan, M. "At the Moment of Sputnik the Planet Became a Global Theater in Which There Are No Spectators but Only Actors." *Journal of Communication.* no. 1 (1973): 48–58.

Metcalfe, R. M. "Wirefree RadioMail to Change My Life, in 1994." *InfoWorld.* November 29, 1993a, 15: 56.

Metcalfe, R.M. "Wireless Computing Will Flop—Permanently." *InfoWorld.* August 16, 1993b, 15: 48.

Milutis, J. Superflux of Sky. *Cabinet. no.* 10 (2003). Accessed January 22, 2012. http://cabinetmagazine.org/issues/10/superflux.php

Morus, I.R. "'The Nervous System of Britain': Space, Time and the Electric Telegraph in the Victorian Age." *The British Journal for the History of Science.* no. 4 (2000): 455–475.

Negroponte, N. *Being Digital.* New York: Alfred A. Knopf, 1995.

Nye, D.E. *Electrifying America: Social Meanings of a New Technology, 1880–1940.* Cambridge, MA: MIT Press, 1990.

Peters, J.D. *Speaking Into the Air. A History of the Idea of Communication.* Chicago: University of Chicago Press, 1999.

Poliakine, R. "About Powermat—Our Story." Accessed June 22, 2011. www.powermat.com/about-powermat

PowerBeam Inc. "PowerBeam: Wireless Power, Wireless Energy, Wireless Electricity."
    Accessed June 22, 2011. www.powerbeaminc.com/technology.php
"Power Without Wires." *The Globe*. April 26, 1934, 1.
Price, C. "What's Wanted from Inventors." *Popular Science Monthly*, 1925.
Schatz, D. Interview by Ghislain Thibault (Watertown, Mass. April 12). (Vice President of Sales & Business Development, WiTricity), 2011.
Stolow, J. "The Spiritual Nervous System: Reflections on a Magnetic Cord Designed for Spirit Communication." In *Deus in Machina: Religion, Technology, and the Things In Between*, edited by J. Stolow, 83–116. New York: Fordham University Press, 2013.
Taylor, C. *Modern Social Imaginaries*. Durham: Duke University Press, 2004.
Tesla, N. U.S. Patent No. 645 576. Washington: U.S. Patent and Trademark Office, 1897.
Tesla, N. "On Electricity: Address on the Occasion of the Commemoration of the Introduction of Niagara Falls Power to Buffalo at the Ellicott Club (12, Jan)." *Electrical Review* (1897): 46–47, reprinted in Vojin Popovialc, Radoslav Horvat, Branimir Jovanovialc, Dejan Bajialc, Miroslav Benišek, et al. (eds.). "Lectures," 274. Belgrade: Zavod za udžbenika i nastavna sredstva, 1999.
Toynbee, A. J. *A Study of History*. London: Oxford University Press, 1934.
Wells, H. G. *Anticipations: Of the Reaction of Mechanical and Scientific Progress Upon Human Life and Thought*. London: Chapman and Hall, 1902.
Yurick Sol. *Behold Metatron, the Recording Angel*. Los Angeles: Semiotext, 1988.

6 The Rise, Fall and Future of BlackBerry™ Capitalism

Andrew Herman and Vincent Manzerolle

INTRODUCTION: "POTUS KEEPS HIS PRECIOUS" STILL

The original inspiration for this chapter was a story about Barack Obama and his BlackBerry. In the run-up to Obama's inauguration as the forty-fourth President of the United States (POTUS), there was considerable coverage in the global media concerning the fate of his beloved BlackBerry. One of the hallmarks of Obama's triumphant political persona at the time was his 'street cred' among the digerati as a savvy Internet user. His reliance on the iconic BlackBerry mobile communications device in the conduct of his successful presidential campaign became legendary (Clifford 2009). So it was with great tribulation that the world waited while it was decided whether or not Obama could keep his BlackBerry in light of the security requirements of his new position. In the February 2, 2009, issue of *Newsweek* magazine it was announced that, yes indeed, "POTUS Keeps His Precious" (Bailey 2009).

At the time, BlackBerry was the dominant smartphone brand and the company which made it—the Waterloo, Ontario–based Research in Motion or RIM—was one of the most successful players in the mobile communications industry. The hegemony of the BlackBerry as mobile communications device and status object of desire was so complete that it compelled one chronicler of its rise to dub its dominion "BlackBerry Planet" (Sweeny 2009). Like Gollum's beloved One Ring in the *Lord of the Rings*, POTUS's "precious" operated as a sublime technological object of desire and transcendent power (Mosco 2005; Nye 1996). But rather than binding all of Middle Earth in fealty to the holder of the One Ring, the BlackBerry offered the technological sublime of ubiquitous connectivity of each to (potentially) all, and so it appeared at the time that Obama's "precious" was a perfect metonym for the emergent spirit of early-twenty-first century information-based capitalism, the spirit of what the first author originally termed Black-Berry Capitalism.

How times have changed. At the time of Obama's public agonizing about being able to keep his "Precious," the stock of Research in Motion was trading on NASDAQ at $67, its global market share for handsets was

20 percent, and its market valuation was $22.5B. By January 2014, RIM had been rebranded as the BlackBerry corporation, its stock was trading at $9, its global market share had declined to less than 1.7 percent, and its market valuation was $3.9B.[1] We have to admit that there is a certain conceit to naming a particular instantiation of a mode of production (capitalism)—one that has lasted for over five hundred years—after a device, a company and a brand. It is a conceit that is begging for comeuppance, particularly when the fortunes of all three fell dramatically and precipitously, consigning our analysis to the dustbin of dead media and corporate history. Who, for example, now remembers the Digital Electronic Corporation and their VAX mainframes that ruled computing until the late eighties? The vagaries of history not withstanding, however, we will argue in this chapter that the concept of "BlackBerry™ Capitalism" is indeed a worthy and insightful heuristic for understanding how the materialities and imaginaries of the mobile Internet comprise a particular assemblage of a capital/ism.[2] A close and critical examination of the rise, decline and continued relevance of the BlackBerry as device, corporate entity and brand will illuminate the articulations between different dimensions of the present epoch of capital/ism, a moment where capital is simultaneously digital, informational and connexionist. Besides, at the time of this writing (March 2014) Obama still cherishes and keeps his precious BlackBerry and will do so, apparently, until it is "pried from his hands" (Star Staff 2013).

The argument of this chapter unfolds in two parts. In the first part, we consider two distinct yet imbricated conceptualizations of the current configuration of capitalism: "digital capitalism" and "informational capitalism." Each conceptualization offers valuable yet partial insight into the materialities and imaginaries of the mobile Internet in general, and what we will ultimately call the BlackBerry™ Capitalism assemblage, in particular. Yet the terms are often used interchangeably in critical media and communication studies, diminishing each one's distinctive analytical purchase of a better understanding of capitalism. This chapter attempts to clarify the terms "digital capitalism" and "informational capitalism," thus enhancing their critical value. In the second part of the chapter, we turn to the multiple ontologies of (the) BlackBerry—as wireless telephony, as Internet portal, as ensemble of socio-technical affordances, as material object, as corporate entity, as commodity, and as brand image—that comprise a specific assemblage that embodies and informs particular articulations of digital and informational capital/ism.

DISCERNING BLACKBERRY™ CAPITALISM

Theorists of "digital capitalism" are rooted firmly in the tradition of the political economy of communication and view the present era as constituting a "phase change in the 500-year history" of capitalism.[3] From within the

historical perspective of the *longue durée* of capitalist development, these
theorists argue that the "telecommunications, media and technology" sector
have "become a leading component of the spatio-temporal fix in which capital attempted to extricate itself" from its recurring systemic contradictions,
most recently manifested in the from the prolonged crisis of accumulation
on the late sixties and early seventies (Chakravartty and Schiller 2010, 672).
The specific social and economic policies that comprise the foundation of
what has been termed "actually existing neoliberalism" (Peck 2012), as well
as the ideological lineaments of legitimating belief in the project of global
neoliberalism itself, have been made possible by the rapid diffusion of information technology and digital telecommunication networks (Fisher 2010).
The financial sector and large corporations invested heavily in digital information technologies as part of shift from Fordist to post-Fordist logics of
capital accumulation in all spheres of the commodity circuit (production,
circulation and consumption) on a global scale (Harvey 1990, 2010). As
Chakravartty and Schiller argue, "Digital capitalism coalesced, therefore,
not as a sectoral or communications centric phenomenon, but as in inclu-
sive, economy wide project" (2010, 672). This reconfiguration of the circuit
of capital is important in understanding BlackBerry capitalism, a point that
will be underscored in the next section of the chapter.

Schiller (2011) identified five specific "vectors" along which ICTs helped
to cast "digital capitalism" into being on a global scale. The first and perhaps most salient vector was the increasing "financialization" of advanced
capitalist economies. National and international financial sectors were themselves the locus of massive amounts of investment in ICTs that provided the
infrastructural basis for coordination and management of the global flow of
capital as well as the development of new financial instruments of securiti-
zation that drove the financial crisis of 2008. The second vector was the military Keynesianism of the U.S. government and its enthusiastic embrace of
ICTs and digital media in the development of the "network-centric warfare"
paradigm. The third and fourth vectors were private corporate investment
in ICTs to reduce labor costs through automation and the management of
global supply chains, respectively. The fifth and final vector is what Schiller termed the "accelerated commodification" of consumer culture, which
refers vaguely and generally to almost every important change and development in the media industry enabled by the Internet (2011, 927–930). One
of the primary consequences of these developments from the point of view
of "digital capitalism" is the concentration/consolidation of market power
in fewer and fewer corporate entities in the telecommunication sector generally and in terms of mobile communications specifically.

From our point of view, the "digital capitalism" analytic is crucial to
understanding the development of the mobile market (comprising hardware, software and services) as being integral to the current historical phase
of capitalism. The adherents of the "digital capitalism" perspective are
quite right to argue that the predominance of digital-based information and

information technology in this epoch of neoliberalism *does not* entail a transcendence of class relations and class power as theorists of the "information society" have been wont to argue (cf. Toffler 1984; Webster 2006). Moreover, their focus on how new digital information technologies made possible the transnational coordination of a global neoliberal economic order is also absolutely correct. However, one of the limitations of conceptualizing the mobile market through the lens of "digital capitalism" is that there is scant attention paid to the specific characteristics of "the digital" as a specific form of media materiality, "information" as the content of that media, and either as being constitutive of distinctive communicative practices specific to the epoch of *digital* capitalism. To put the issue another way, what is needed is a consideration of the ways in which specifically *digital* information technologies—particularly when they are part of the mobile Internet—constitute both the process and the product of capital accumulation. In fact, after taking great pains to distinguish their analysis of digital capitalism from theories of "the information society," Chakravartty and Schiller (2010), for example, ignore the question of what, if anything, is particular about information in digital form in our own era that might distinguish it from other (predigital) forms of information in the five-hundred-year history of capitalism.

This aporia in the analytic of digital capitalism, we believe, can be partially addressed by a turn to materialist medium theory.[4] As Canadian medium theorist Harold Innis (2008) argued, the development of media forms and communicative practices has been inextricably bound up with commerce since the beginning of formalized writing practices in the ancient civilizations of the 'fertile crescent.' In the earliest written texts to emerge from this region, communicative practices are given material shape—articulated so that market transactions/relations could be represented, imagined, and constituted over space and in time. Innis, for example, argued that the widespread use of paper and ink at the beginning of capitalism's five-hundred-year reign enabled the cartographic mapping of political and commercial empires, connecting the frontiers of empire with the metropolitan "backtier," and enabling control over the flows of capital between the two. Inspired by Innis's analysis of the spatial and temporal biases of different mediums, James Carey (2008) argued that diffusion of the assemblage of telegraphy across North America was essential in creating national and international futures markets for agricultural commodities where price became disembedded from place and traded against quanta of time.

What both instances of media development represent is a fundamental aspect of how the materialities of media forms engender(ed.) new capitalist imaginaries: the abstraction of knowledge from lived experience and its transmogrification into "information" creates the fiction of a readily exploitable "new" world ripe for the taking as empire and re-making as capital (cf. Headrick 2002; Steinberg 2005). As James Beniger (1986) argued in his now classic work *The Control Revolution*, "information derives from the organization of the material world" (9). Pushing this point in a Foucauldian direction, we might say that information makes the pragmatics of power of organizing the material world possible (Foucault 1980). The production, dissemination and storage of information has been central to the capitalist enterprise ever since the Renaissance.[5]

The value of integrating the materialism of the political economy approach with the materialism of medium theory[6] is that this integration highlights both the relationship between media forms, their socio-technical affordances, and the communicative practices they enable, and how these articulations produce regimes of power-knowledge that are intrinsic to new forms of capital and its accumulation, on the other. The point here is that "information"—as a mode of control and coordination—and its media/tion is always central to the constitution of capital and the operations of capitalism.

Interestingly, it is in precisely how information is *digitally* constituted as a mode of coordination and control that distinguishes theories of informational capital/ism from those of digital capital/ism.[7] The theories and theorists of informational capital/ism are quite heterogeneous in their conceptual rendering of this epoch of capitalism, yet they all share a common understanding of the profound changes wrought by, first, the technical character of the digital materiality of information and second, the socio-technical organization of its production and distribution, Andreas Wittel (2012), for example, argues that is the articulation of *digital media forms* with *distributed media practices* that comprise the fundamental difference between, to use Benkler's (2006) terms, the "industrial information economy" of the twentieth century and the "network information economy" of the twenty-first century. Wittel identifies four characteristics of digital media that create a unique set of socio-technical affordances of distributed media with the possibility to reshape relationships of power:

> They can re-mediate older media forms such as text, sound, image and moving images as digital code; (2) they can integrate communication and information, or communication media (the letter, the telephone) with mass media (radio, television, newspaper); (3) digital objects can endlessly be reproduced at minimum costs; (4) they don't carry any weight, thus they can be distributed at the speed of light.
>
> (Wittel 2012, 317)

Optimistic theorists such as Benkler (2006) and Cortada (2011) argue that the technological infrastructure of the "networked information economy"—an infrastructure that is based upon media that is digital *and* distributed—radically reconfigures relations of power as the production, distribution and consumption of mediated communication move from the hierarchical to the flattened, from the proprietary to the commons, and from the inflexibly mass to the flexibly peer-to-peer. The production of cultural meaning and the flow of information are decentralized to the point where the dominant institutions and forms of mass communication are increasingly displaced and marginalized

However, most critical theorists of informational capital/ism do not share Benkler's optimism of the will of the peer-to-peer commons. Although there is no question that the development and diffusion of "Web 2.0" applications and platforms have significantly transformed the landscape of media and cultural production, there are other elements of informational capital/ism that mitigate their democratic potential. The very qualities of the digital media forms that have radically reduced the costs of production and distribution, namely, their "immateriality" as code, have engendered tendencies that enriched and empower what McKenzie Wark (2012) terms the new "vectoral class" of financiers and entrepreneurs who own and control the key Internet and social media corporations.

Crucial elements, or characteristics, of the informational capital/ism paradigm include: (1) money (including credit) is itself only nominally material but highly informational, leading to a deepening interdependency/ integration between financial and ICT industries; (2) the rise in "immaterial wealth" production that is comprised of media creations and artifacts that are increasingly defined as intellectual property by media corporations as well as financial valuations of corporate brand identity constituted by goodwill; and (3) much of this "immaterial wealth" is user-generated or produced by produsers/prosumers and their "immaterial labour," that is appropriated by corporations as intellectual property (IP).[8]

The creation of immaterial wealth through immaterial labor is increasingly dependent upon material assemblages. Yet because the tendency under informational capitalism is toward "immateriality," the formal and material aspects are often overlooked or disregarded. Our task in the next section is to highlight the specificity, indeed, *materiality*, of the accumulation process (e.g., the process by which fixed and variable capital are organized for surplus value production and accumulation). This requires exposing the sinews (e.g., media) that materially bind together capital's increasingly "immaterial" corpus (e.g., "the virtual corporation").[9]

These sinews, however, are themselves highly capitalized (and commodified), leading to a competition among branded ICT "solutions." This competition— to capitalize and brand the essential infrastructure of capital accumulation and thus to profit from the essential tools of "competitive advantage"—leads to regular intervals of volatility and "disruption" unleashed by the "gales of creative destruction" (Schumpeter 2008). Arguably, the rise and fall of the BlackBerry is evidence of precisely this process so central to the ideology of free and competitive markets (e.g., Neoliberalism). BlackBerry™ Capitalism therefore reflects not only a particular material assemblage articulating the use of the mobile Internet to instantiate the virtual corporation, but also highlights the competitive logic which contributes to the specifically *branded* nature of this assemblage.

THE ASSEMBLAGE OF BLACKBERRY CAPITALISM

BlackBerry Capitalism is a branded iteration of capitalism in its 'informational' mode, that is, it offers an 'end-to-end' system which seamlessly and ubiquitously networks the individual to capital's organizational structures. In this case we can think of the progressively 'virtual' nature of corporations and workforce management. Thus BlackBerry offers three ways by which this networking is both materialized and itself capitalized upon: ubiquitous connectivity (UC), virtualization (via new registers of mobility) and prosumption. Moreover, these offer a means of understanding the relationship between BlackBerry Capitalism, and the forms of "social mobilization" captured by the term "informational capitalism."

Ubiquitous Connectivity

As a "media condition," ubiquitous connectivity can be usefully analyzed in terms of how material media forms express a convergence of ubiquity, immediacy and personalization in their habitual uses, composition and commercialization. These features are characteristic of informational capitalism generally, and in BlackBerry Capitalism specifically.

Ubiquity here refers to both the perceived and actual colonization of digital media devices and, in this case, the technical capacity to remain connected at all times in devices designed to be "always on" and "always on you." This experience is framed by the user's perception or desire for immediacy in the production and consumption of data, and the personalization of devices themselves.

Immediacy refers to a perceived instantaneity (or simultaneity) enabled by the devices and infrastructure, tending toward real-time, networked communication and a logistical collapsing of spatial distance. Connectivity, comprised primarily of both the transmission and reception of digital data, is relatively unencumbered by spatial and temporal constraints, effectively tied to the specific location of individuals.

Personalization refers to the tendency of contemporary media to materially incorporate the identity, information and relationships of a particular user. The identity of the user is integrated into the commercial development of digital media as well as in its technical composition (e.g., SIM cards, NFC chips, unique device identifiers). Personalization of digital media also refers to the filtering (Pariser 2011) and the customization (Turow 2011) of content and services, for example, through the embedding of algorithms that learn the habits of particular users (Mager 2012).

Although each of these aspects on their own is not entirely new or unprecedented, what is new is the scale of their configuration in the myth-making activities associated with a specific media artifact (the BlackBerry) and, more broadly, as the combined appearance of a relatively new category of consumer technologies and services. Thesse comprise the assemblage of

BlackBerry capitalism. Thus we focus on the BlackBerry as a case study because it offers an ideal media artifact with which to historicize both the imaginary dimensions *and* technological appearance of informational capitalism (see also Manzerolle 2013).

New Registers of Mobility and the Virtualization of Capital

Social mobilization under informational capitalism entails the deployment of ICTs to manage the dispersion and movement of a largely "virtual" workforces. Yet this digital labor, as materialized in the artifact itself, is representative of a particular condition of mediation linking these always-on devices with their always ready to work users. The development of BlackBerry Capitalism also relied on a concurrent "re-imagination" of the commercial enterprise itself. This new corporate imaginary emerged from a burgeoning management literature emphasizing both the "virtual" workforce and the "virtual organization." New devices and services are required in this spatially and temporally dispersed workforce to maximize both connectivity and flexibility, fitting with the project oriented nature within the "new spirit of capitalism" (Boltanski and Chiapello 2007).

Management literature heralding the rise of the *mobile workforce* extended the telework discourse to stress the connectivity of workers, whether by transportation or networked ICTs (Pratt 1997). Indeed, the early proselytizing around wireless data networks and services emphasized the importance for mobile workforces (Brodsky 1990; Didner 1991; Hengel 1994; Ryan 1991). The logical conclusion of the mobile workforce is to make them available 'anywhere and anytime.' A 1994 article in *Industry Week* proclaimed the coming "anytime, anyplace workplace" encompassing "an ever more mobile workforce that is connected by technology" and "influencing the scope of work and how we get tasks done" (Verespej 1994).

Flexibility is the watchword underlying both telework and the mobile workforce, and this is what makes each such a potent example of how the imaginary of capitalism is materialized in technologies, practices and ways of thinking. Through the application of ICTs, flexibility comes to define both capital accumulation and labor management. Thus, in the myths of post-industrialism, flexibility is synonymous with connectivity. At the same time, such flexibility also conceals the pervasive managerial surveillance and control implicit in the application of telework (Fitzpatrick 2002; Huws et al. 1990; Lyon 1994; Newitz 2006).

The "Always On" Prosumer

Social mobilization under conditions of ubiquitous connectivity creates a particularly new spatio-temporal dispensation of communication and production (Herman 2013). More specifically, this can be seen in the development of social media into platforms for incorporating and capitalizing upon activities of prosumption. This entails not only the exploitation of user-driven content, but also of the valuable metadata created automatically as a result of any interaction with the platform. Prosumption is important to the broader reproduction of informational capitalism for these two interrelated reasons: it progressively mediates the expression of communicative and creative capacities of users (e.g., content), while also capitalizing on the valuable metadata that can be directly incorporated (licensed or sold) into the systems of production and distribution. Indeed, the "circuits" of capital are prospectively sutured directly into the spatio-temporal habitus of everyday life via UC as a participatory media condition.

The economic necessity of information to contemporary capitalism has contributed to the renewed popularity of the prosumer—a figure that, since its popularization by Toffler (1980), embodies the convergence of production and consumption within the purview of an empowered and autonomous user-consumer of ICTs (see Bruns 2008; Comor 2011). The prosumer, however, is in fact the techno-utopian representation of the sovereign consumer championed by neoclassical economists (Gowdy and Walton 2003). In accordance with neoliberal theory, this figure provides a digitalized version of human rationality premised on self-interest. Thus it is thus not surprising that Web 2.0 reflects a kind of neoliberal form of individualism that posits consumer sovereignty in the creation of user-generated content—a symbol of the empowerment of rational individuals over networks. Informational capitalism privileges the commercial development of the vast information supply chain that depends on the real-time participation of the prosumer-user, but also the regular supply of metadata about user behaviour for a whole range of economic actors (app developers, hardware manufacturers, telecoms and government agencies of various kinds).

BLACKBERRY CAPITALISM: MATERIALITIES AND IMAGINARIES

The analytic significance of the BlackBerry is not intended to offer praise for its technical or commercial achievements, but instead to show how these achievements express a synthesis representing the motivations of economic actors and prevailing modes of thought as they are drawn together in and through a particular, branded instantiation of informational capitalism.

Given its waning commercial standing, it is easy to forget that the BlackBerry brand, for a time, inspired fanatical devotion, spawning various online fan, user and support groups (Michaluk 2011). The fanatical devotion to the BlackBerry brand, and its UC-enabled lifestyle, was rivaled only by those initiated into the "Cult of Mac." As evidence, consider that in 2001 a *USA Today* cover story declared the BlackBerry to be "the heroin of mobile computing" (Maney 2001). Yet despite what many of the manic technology "experts" may have said, the BlackBerry's significance is not in its singular existence as a popular product, brand or investment stock. Nor is its cultural, technica, or political-economic importance. Rather, it is in the way that these all come to be tied together in a specific socio-technical assemblage

The BlackBerry, as we have said, is an ideal artifact of analysis for engaging with the imaginary of informational capitalism. In the earliest stages, this specific brand of capitalism was defined primarily by ubiquitous email, the push-based communication of data wirelessly and "always on" functionality. The BlackBerry brand, as this chapter demonstrates, evolved to pitch ubiquitous connectivity as a part of sociality itself, an essential mediator of our identity and social networks—a sentiment given form in the 2010 BlackBerry slogan "take life with you." The BlackBerry also affords an opportunity to see the constitution of a generally new category of devices and services built on infrastructure unique to the mobile Internet (e.g., packet-switched mobile standards like Mobitex, ARDIS networks). Thus if we line up every BlackBerry model, from first to most recent, we see expressed in them the commercial impulses that have now seemingly enveloped the entire globe: the construction of UC not simply as a condition of work consonant with the networked organization or economy, but of a new lifestyle that reflects both a series of ruptures and continuities with the cybernetic imaginary of capitalism itself. This is a story that is visually told in the physical evolution of the BlackBerry; that is, its morphology as a consumer device, outwardly expressed in a way not possessed by later technologies like the iPhone, in which one observes only relatively small outward differences among generations.[10] However, the story is not only one of outer appearance, but also of internal operations, software, processing power and miniaturization. As a case study, the evolution of the BlackBerry illustrates the colonization[11] of everyday life by computer processing, arguably reaching a developmental finality begun with the popular deployment of the transistor, followed by the integrated circuit and finally the microchip.

This imaginary is embedded in the recomposition of the labor force away from comparatively inflexible unions and other contractual arrangement toward more flexible (and precarious) forms of work. As myth is made *real* through new devices and services, BlackBerry capitalism builds on a vision of a future workforce and culture, one requiring new tools to meet and manage their professional and social needs in order to make them more flexible, manageable and productive. As such, new techniques and conceptualizations emerge as the adoption of devices spreads. BlackBerry capitalism stems from a belief in a future *networked* worker whose professional and social lives are inseparably intertwined.

Device/Artifact

> One of the things we found was that the closer that these devices get to you physically, if it's something that you wear, like a watch or something you put in your pocket or hold in your hand, it takes on a whole

different meaning. It becomes something that becomes a part of you. It reflects your character, your personality, and you take certain attributes from that device. There's a status involved. So what we found was that there's going to be a large amount of customization going forward. There are going to be different styles, different models.

(Lazaridis 2008, 12)

The materiality of BlackBerry Capitalism begins most tangibly with the commodity object, the device, which is the interface, or relay, between the situated actor and capital itself as represented by a particular organization (whether commercial or noncommercial). It is the artifact of BBC that allows us to start a deeper exploration of the particular conditions of mediation that underpin both BBC specifically, but also informational capitalism more generally. It is also the interface point between the individual and broader capital (infra)structures (base stations, servers, fiber-optic cables, satellites, etc.). The device is not only the most vivid commodity object standing in for the much broader infrastructure (materiality), but it also represents a particular tool for 'digital labor.' Here digital labor can be construed generally as those activities that contribute to a net increase in the 'input(s)' of digital data into capital's circuit of (re)production.

What is specific in the case of BlackBerry Capitalism is the emphasis on precisely the "digital" entry of information via the QWERTY keyboard operated by the thumbs. Indeed, the device was conceptualized from the outset as a platform for the production and consumption of information, navigated via the thumbs (thus creating a "thumb culture"; see Glotz et al. 2005).

Another distinguishing feature of the BlackBerry that separated it from popular rivals like the PalmPilot was the choice of interface—touch screen vs. QWERTY—which allowed the BlackBerry to emphasize email as its major functionality with peripheral PDA-like capabilities appended. Although emphasis on email fits into the pragmatic and personalized simplicity of wireless data, it was the divergent application of interface technologies that allowed for its core differentiation. The choice to use QWERTY was built on existing familiarities with then existing technology, protocol, competency and other tacit forms of knowledge. Moreover, the choice of the full QWERTY keyboard made the parallels to stationary computers clearer to consumers. In this respect the keyboard emphasizes the active and productive side of wireless communication.

In combination with the emphasis on email functionality, the choice of QWERTY is an important transitional precursor to the era of ubiquitous connectivity; an era in which users are both producers and consumers of mobile data. One can cite here McLuhan's observation that the content of all new media is a previous medium (McLuhan 1964, 23) as portable computing, and UC with it, required QWERTY as a bridge to acceptance. The importance of QWERTY as a transitional interface built on the existing competencies of users was central to enabling and equipping people to become accustomed to consuming and producing data ubiquitously. It therefore makes sense that the keyboard for RIM would be an important piece of intellectual property as well as a key component of the BlackBerry brand identity.

As the primary data entry point for mobile devices, BlackBerry's QWERTY maximized communicative efficiency and ease of use through its optimization of thumb typing. Fortunati (2005) has pointed out how mobile devices represent artifacts of what she calls a "thumb culture" as a distinct set of cultural practices, forms and relations stemming from the use of devices controlled by the thumbs (149–160). Although this characterization certainly reflects the prominence of text-based communication enabled by the QWERTY keyboard, "thumb culture" and "digital labor" are distinctive characteristics of the era of UC. Combined with the global popularity of texting in the wake of GSM standardization, the BlackBerry made such activity a basic element of the workday, but one whose use value was easily translatable to the social lives of users already familiar with desktop computer keyboards.

Branding

The first annual report after RIM's initial public offering (IPO) in 1997 offered its corporate narrative for its product offerings—a narrative crafted to reflect the company's indispensability to consumers and the telecom providers that would profit from the devices monetizing their network capacities. The message was not overly complex, reflecting the influence of ICTs on the rhythms of everyday life. Communicating to investors, employees and prospective consumers the company's newly established corporate narrative, the report proclaims, "In a world where consumers increasingly demand to be 'connected' 24-hours-a-day for both business and personal purposes, the economical cost and benefits of RIM's core two-way paging technology give it a significant competitive advantage over the limited applications of one-way products which cannot respond to or initiate messages" (RIM 1998, 4). Not only was the device positioned as one designed for UC, it also was portrayed as a device for both consuming *and* producing wireless data.

The next step was to develop a brand identity that reflected the unique "value proposition" (i.e., use value) offered by RIM's new device—one meshed with the future needs of wireless telecommunications providers. After some consultation with branding experts and product designers (Colpatino 2011), the name "BlackBerry" was adopted in 1999 as a means of distinguishing the device from others existing in the market. The name, rigorously tested in focus groups by Lexicon Branding,[12] was partly chosen as an ideographic reference to how the QWERTY keyboard of the device resembled the fruit, and partly chosen because linguistics research suggested the appeal of the double "B" sound to focus group participants (MacNamara 2012). The BlackBerry brand was to be all-encompassing: it was simultaneously a device, a complex technological system and a service enabling individuals (whether as professionals, entrepreneurs or as part of the larger workforce) to remain connected at all times. While the brand was associated to the physical device, BlackBerry represented, symbolically, a *service* that provided and personified UC (RIM 2001). Though the devices themselves were expected to evolve, growing in technical and functional sophistication over time, the basic service and its connotations could remain the same, indefinitely tied to the BlackBerry brand name.

No longer a pager *per se*, the newly branded BlackBerry850/950 was released in January 1999 on multiple carrier networks in North America. One of the most important selling points was the ability to *push* data to users so that there was not lag time between when users are notified about a new message and when they open that message. The push-based system RIM had been working on for years became a central and defining feature of the "BlackBerry Solution," allowing personal information to find its designated receivers regardless of where they were in space. Arguably, it is this particular technological innovation that made UC a question of perceived immediacy (since there was virtually no gap between getting the notification of new email and actually accessing that email). Indeed, the ability for the BlackBerry to be used almost anywhere in space, and then to use that location to push wireless data, set the stage not only for the mobile Internet but also for an existential condition in which bursts (and flows) of information can interrupt daily life virtually everywhere and at anytime. The slogan, "always on, always connected," was part of an extensive public discourse on UC that shaped RIM's corporate identity as well as the bourgeoning PDA market (Wasserman 2001).

The marketing campaign for the BlackBerry957 was the first concerted campaign led by RIM and not by carriers (RIM 2001). The campaign captured the essence of the BlackBerry brand, reinforcing a narrative that not only valorized "always on" connectivity but also depicted the device as a necessary tool suited to the Internet-age of global ICT markets—a mythos in which the speed of information was a defining characteristic, requiring new tools of adaptation. A series of ads circulated in 2000 and 2001 appearing in Canada's *The Globe and Mail* are paradigmatic of this overall narrative. One advertisement depicts a man on a golf course checking his email, while another shows a woman lost at sea in a rubber dingy presumably sending a distress message with her 957. More telling still is an advertisement that depicts a man narrowly avoiding a knife thrown by some unknown assailant; the message on the BlackBerry screen simply reads "DUCK!"

Advertisements during this campaign often alluded to, and made light of, the addictive nature of wireless email. For example "It comes with an off button. No one uses it, but it comes with one"; "If you're planning an intervention for someone addicted to one, may we suggest you use e-mail";

"You shouldn't use it in the shower. You'd think we wouldn't have to say that." Each of these ads describes the BlackBerry as a device with "highly addictive wireless e-mail."

Platform OS (Mobile Internet)

This new field of devices and services is itself an attempt to commercialize the long-touted "mobile Internet," which, unlike its wireline predecessor, was subject more closely to the machinations of market competition (and the negative effects of standardization, patents and incumbent telecom interests). BlackBerry capitalism is a particular development of the mobile Internet as a platform for broader wireless services to develop. Of specific importance is the development and commercialization of "push-based" communication through the use of wireless packet-switching infrastructure which allows messages to "find individuals" in time and space, enabling the potential for a state of ubiquitous connectivity. The Internet boom catalyzed further interest and investment in wireless data networks culminating in the marketing of 3G as the grand arrival of the truly *mobile* Internet (Edwards 1998).

The Virtual Organization

At the organizational level, Castells (1996) declared the rise of the "network enterprise" as a key transformation in the overall mode of production of contemporary capitalism (187). This rhetorical emphasis on networks was echoed in the business press and by management consultants (Baker 1994; Tapscott 1996), all suggesting a fundamentally altered corporate structure more properly aligned with the visions and values of other post-industrial discourses (like telework). For corporations this meant being able to utilize the 'sunk' investments represented in fixed capital, contributing to more 'flexible' and 'decentralized' *networked* information technologies as mediators of innovation and productive efficiency. It is therefore not surprising that wireless data technologies (and with it, UC) coincided with the rhetoric of "virtualization" to describe the application of network technologies in reshaping the spatial and temporal organization of the business enterprise.

Virtualization contributed to an existing discourses surrounding telework a more ephemeral, yet totalizing, description of organizational forms and labor processes. This emergent understanding of *virtual* as having effect but not form directly preceded the era of UC. It realized this definition by extending the power of management over workers through information flows to and from workers regardless, in theory at least, of the position of employees in time and space. Perhaps more tangibly, virtualization can be used to describe new forms of commercial resources: assets (fixed costs are substituted for variable costs), employees (those that do not need to be physically located in a centralized office) and time ("resources of time seem to expand or shrink at will") (Birchall and Lyons 1995, 18). As Morgan (1993) described, in the early years of the Internet, virtualization was seen to constitute a major shift in management's vision of both the organization of commercial activities generally and labor specifically:

> Organisations used to be places. They used to be things . . . But, as information technology catapults us into the reality of an Einsteinian world where old structures and forms of organization dissolve and at times become almost invisible, the old approach no longer works. Through the use of telephone, face, electronic mail, computers, video, and other information technology, people and their organizations are becoming disembodied. *They can act as if they are completely connected while remaining far apart. They can have an instantaneous global presence. They can transcend barriers of time and space, continually creating and re-creating themselves through changing networks of interconnection based on 'real time' communication.* . . .the reality of our Einsteinian world is that, often, organizations don't have to be organizations any more!
>
> (Morgan 1993, 5,emphasis added)

The concepts of virtual work, virtual teams and virtual organizations proved popular enough with business strategists and management experts to spawn numerous "how to" manuals, guiding management professionals on how to implement virtual strategies in their own organizations. Consequently, theories of the "virtual organization" (Quinn 1992; Davidow and Malone 1992; Mowshowitz 1994; Birchall and Lyons 1995; Grenier and Metes 1995; Fukuyama et al. 1997; Jackson and Wielen 1998), "virtual work" (Jackson 1999; Watson-Manheim et al. 2002), and "virtual teams" (Ebrahim et al. 2009) have become (and remain) popular in publications addressing the business impact of new ICTs. The proposed benefits are familiar truisms for business literature: efficiency and productivity gains benefit management, while increasing flexibility and empowerment benefit workers. Paul Drucker and his followers even alluded to the "virtues of virtuality" years before this virtual thematic began to appear in management literature *en masse* in the 1990s (see Micklethwaith and Wooldridge 1996, 112–114; Hesselbein et al. 1997, 377–383).

The evolution of telework literature toward virtualization provided a ready climate for the introduction of wireless devices for labor management. Indeed, this is precisely what RIM's BlackBerry Enterprise Server (and related software applications) addressed. Management experts had provided a set of problems; RIM provided the technical fix for the drive toward virtualization. The BlackBerry enterprise server was one of the first technological systems that directly facilitated the "virtual organization" by wirelessly tethering networked connectivity to the worker (Dewar 2006). As it became an early "killer app" for the mobile Internet by, in effect, colonizing the enterprise (Harmon 2000), wireless email, was also, in effect, a "Trojan horse" (Maney 2001) socializing workers into accepting the "condition of immediacy" (Tomlinson 2007) enabled by UC.

As elite business users began to adopt BlackBerrys in growing numbers, RIM poured more effort into developing and marketing its BlackBerry Enterprise Server (BES). While the BlackBerry was most visibly expressed as a singular handheld device, the BES was marketed as an integrated technological assemblage comprising end-user devices, network servers, back-end support, software and hardware; in sum, it was a "total package" that allowed any corporate client to implement a secure wireless strategy with relative ease and speed. The BES enabled total data and network synchronization across a mobile workforce. It allowed for a level of customization according to the client organization's information, security and networking needs. BES therefore forged not only a large potential market for RIM to exploit, but also solidified the economic necessity of wireless strategies for competitive advantage (RIM 2000). At the end of February 2001, 2,800 companies in North America were using BES (RIM 2001, 10), and in 2003 the number of these servers that were installed by corporations globally exceeded 10,000 (RIM 2003, 4). By 2005 the number reached 42,000 (RIM 2005, 6).

Ilkka Arminen, professor and researcher at the University of Helsinki, makes a case echoing these issues in the era of UC; "Mobile communication anytime, anywhere, increases social accountability. The revival of 'dead' moments not only gives us extra time, but also makes us open to real-time monitoring and control. Mobile communication etiquette seems to involve the norms of 'being always available' and 'reciprocating messages/calls you get'" (2009,97). This engenders, he continues, "normative pressure for availability [while it] also allows [for] an increase in accountability, a continuing monitoring of communicative parties" (Arminen 2009, 97).

Archetypes and Service-Affordances

> The BlackBerry freed us. It freed me. It freed others that used the product because it allowed us to leave the office, go home, spend time with the family, and not feel stressed out because you might miss an opportunity, or you might not be able to help out at work when there was a problem and people needed your help. So in effect what it did was it allowed you to get something done very quickly. It allowed you to get it done accurately, and get it done within a short period of time. So you can spend more time with your family, more time with your personal pursuits.
>
> (Lazaridis 2008, 8)

BlackBerry Capitalism is not solely dependent on "physical" goods, goods that require complex chains of production (and related costs), as well as pesky flesh-and-blood workers, creating increasingly razor-thin margins.

Rather, the goal is to develop sources of revenue dependent on "intangibles" like services, rents on intellectual property or copyrights. Thus BBC reflects the tendency to package this new condition of mediation as a "branded experience"; indeed, BBC offers a branded experience of the mobile Internet, crafted to meet the new ICT needs of informational capitalism and its most privileged workers.

Given that there was no clearly established market for the BlackBerry, building a market required identifying early adopters who would not only see the value in RIM's device and service but also help grow its prospective market. The marketing strategy for the BlackBerry first focused on seeding the device with high-profile executives, many coming from the financial industry or Silicon Valley—industries where timely messaging was deemed to be extraordinarily valuable. This strategy was intended to both build word of mouth among elite early adopters and, perhaps more importantly, lead elite users (like CIOs or IT administrators) to pressure larger institutional clients into buying BlackBerrys in bulk to equip their workforce. RIM's executives believed that these elite professionals would clearly see the value in UC as it helped them cope with the chaotic rhythms of the global financial and high-tech markets of the late 1990s. "RIM took a grassroots approach to building brand awareness for BlackBerry. Sales people were dispatched as wireless email evangelists to educate Fortune 1000 companies about the availability of an enterprise-class solution for wireless e-mail" (Elkin 2001).

Industry-specific applications like those in finance, law or information technology led to the professionals in these industries becoming the fastest adopters of the BlackBerry (RIM 2000); that is, sectors in which activities, decisions and actions had to be made immediately for the sake of competitiveness and/or profitability. Some early applications for the 850/950 included stock monitoring and trading abilities (RIM 2000). Thus the BlackBerry's core functionality—"always on" connectivity, push email and a QWERTY keyboard—reflected the speed and urgency of timely, round-the-clock flows of information.

While large corporations were sought because of the substantial orders they could place, RIM also targeted entrepreneurs and small-business professionals. In a feature article that appeared in both *The Globe and Mail* and the *Boston Globe* titled "Entrepreneur Grabs Latest Handheld Technology," the popularity of the BlackBerry is explained in terms of how the device empowers such users (Healy 2000). One businessperson is quoted as saying the BlackBerry was his "greatest freedom-provider ever." Another interviewee notes how the device is perfect for venture capitalists because it mirrors their typical "attention deficit disorder" stating the BlackBerry "has totally influenced the way I get business done" (Healy 2000). Indeed, the professional and small-to-medium business market had been an important growth sector for the early development of smartphones. An interviewee describes the importance of the BlackBerry to entrepreneurs and SMB employers and employees: "The BlackBerry is now my watch, my alarm clock, my scheduler, my timetable, my to-do list, my contact list and my internet wireless communication device" (Wintrob 2001). Perhaps more interestingly, the same person describes the wireless feature as "the closest thing to mental telepathy" (Wintrob 2001).[13]

In catering to a more general professional and "creative" class of workers, RIM had crafted a particular artifact that embraced and valorized this "new technological condition" (see Reeves 2007) for an extensive analysis of RIM's promotional discourses). As Reeves (2007) writes,

> the discourse of the devices [the BlackBerry] is reflective of global shift toward a "new economy" ideology that promotes an ethic of productivity and a sense of borderless fluxes. The result for the promotions of the BlackBerry . . . is that the connectivity it enables is presented as a means of increasing productivity. As this new ideology—or ethos— has developed, boundaries [between work and social life] have become increasingly blurred.

Promotional strategy and imagery congealed into a very specific identity for the BlackBerry involving the integration of work and social life. For corporate and business customers, the BlackBerry represented a tool for making the communicative and creative capacities of labor more productive and efficient. For the individual consumer, it was a tool of adaptation to a new technological condition—a condition in which the flows of work and leisure resembled the global flows of information and capital.

The BlackBerry's brand identity stressed the device's ability to remain connected at all times, and to link this ability to an economic and cultural necessity: that competitive advantage, efficiency, productivity and *even social life itself* depended on the individual remaining connected and being able to articulate one's communicative capacities in this way. BlackBerry's brand was precisely about providing this increasingly important ability— constant connectivity—to individual users, organizations and institutions. As such, the brand was a crucial predecessor to the coming age in which the prosumer was no longer a discrete market segment, but a functional social actor, "always on,"[14] performing the role of postindustrial archetype: the prosumer.

RIM's brand messaging and product offerings embraced the prosumer (as both producer/consumer and 'professional user') as its ideal user. RIM's strategy involved reconceptualizing UC—through its devices, services, marketing and investor relations—into a fully connected lifestyle adapted to the new era of empowerment and freedom described by Web 2.0 proponents. A central component of RIM's strategy involved positioning its brand in experiential terms to demonstrate the benefits, and indeed the necessity, of a fully connected lifestyle. RIM's specific goal was to generalize the significance of UC for all as a means of embracing and articulating the archetype of the prosumer (Hamblen 2008).

To accomplish this, RIM began leveraging the iconic aspects of the Black-Berry's brand identity—captured by its first slogan, "always on, always connected"—into a far-reaching message about a radically new social milieu accessible through its devices and services. To do this, RIM developed a new marketing strategy and new consumer-friendly products that tied its identity to the most important elements of Web 2.0. Always on, always connected became as much a social necessity for the Web 2.0 prosumers as it was a business necessity for the virtual organization.

At the height of the Web 2.0 euphoria in 2006, RIM introduced what would become a multi-pronged strategy focusing on the affective qualities of the "BlackBerry experience." This new focus offered an evolved brand narrative bridging work and social life, new devices developed expressly for Web 2.0 prosumers (including added or enhanced media functions) and an emphasis on social networking as a core capability. While appeals to business users focused primarily on access to time-sensitive email, beginning in 2006 a new narrative stressed the affective dimensions of UC: "Love what you do," "Take life with you," "Master your everyday," "Life on Black-Berry." The themes of "love" and "everyday life" are repeatedly deployed in RIM's marketing and advertising beginning at this time.[15]

This broadened narrative shift was also communicated to investors and business analysts in RIM's annual reports, and these constituted what is perhaps the most concise expression of the BlackBerry's expanded brand identity. For purposes of contrast, consider these relatively bland opening lines from the 2006 Annual Report (fiscal 2005), the year before the afore-mentioned transformation:

A World of Information

The flagship product of Research In Motion Limited, BlackBerry is a leading wireless connectivity solution, providing access to a wide range of applications on a variety of wireless devices around the world. It combines award winning devices, software and services to keep mobile professionals globally connected to the people, data and resources that drive their day.

(RIM 2006)

These lines are highly descriptive, factual, and explain clearly the "value proposition" for potential users and customers looking to implement RIM's devices and services as tools of productivity. The BlackBerry is described primarily as a practical wireless business "solution."

The 2007 annual report, however, opens with these telling lines:

Wireless access to email and other information is no longer a luxury reserved for top executives. People everywhere are leading increasingly unwired lifestyles, dynamically balancing careers and rich personal lives.

They need to be able to go where life takes them without losing touch with the people and information that matter most. They need a mobility solution that can blend innovation, usability and style. . .Wireless connectivity is liberating and people who live busy lives want that freedom.

(RIM 2007, emphasis added)

Extending the lifestyle narrative, the 2008 annual report emphasizes UC as a primary selling point to consumers, opening the report with the promise of connecting you to "everything you love in life" (RIM 2008, 2) including social networks, entertainment and leisure activities. The 2008 report goes on to innumerate the various ways the BlackBerry has intervened in everyday life as a necessity, delivering a crucial message to potential consumers and investors alike: RIM is not just about business users, but instead is about a radically new way of life premised on UC in which work life and social life are seamlessly interwoven. This is a particularly important component of the BlackBerry's new expanded identity as it attempts to overcome the potential work-related stigma typically associated with the brand. The 2009 report detailed sections outlining the lifestyle characteristics of the new devices and features, proclaiming that the BlackBerry would "connect to your favorite entertainment," "connect to your social networks," and "connect to your interests" (RIM 2009).

The focus on experiential and affective qualities is an essential part of contemporary marketing and advertising (Arvidsson 2006) and has been a historically important part of wireless telecommunications marketing (Goggin 2006). Focus on experience is arguably more important for wireless services in part because the key medium, the electromagnetic spectrum, is itself experientially intangible. Branding mobile phones and devices as tools of everyday life requires a heavy dose of affect. As Adam Arvidsson writes, "In the case of mobile phones, branding means first of all, the inclusion of customer's everyday life" and in so doing "to construct various forms of branded communities" (Arvidsson 2006, 116, 118).

What is on offer is a service that provides us ubiquitous access to our social lives, positioning it as a basic necessity akin to food and shelter. There is also a political-economic necessity for the wireless industry itself; "As a source of revenue thus shifts from network and call charges to the provision of services and 'content,' the brand also comes to function for investors as a direct indicator of potential future Customer Lifetime Value" (Arvidsson 2006, 116).

With this point in mind, RIM's emphasis on the BlackBerry as a lifestyle necessity refocuses the brand not simply as a specific device but toward the entire BlackBerry brand as an essential service. For RIM, the BlackBerry brand becomes a locus for emotional and affective labor associated with social connectivity. Consequently, the participation (prosumption) of users—particularly younger consumers—becomes a means of adding value to the brand itself as a function of the *network effects* (Benkler 2006)—or perhaps network "affects"—that see the value increase as more people are added to such communities. The more users that are committed to the connectivity offered by BlackBerry, the greater its social and economic value. In this regard, the attraction of younger consumers becomes essential.

THE FALL OF BLACKBERRY™ CAPITALISM

In 2009, the relative success of the BlackBerry as a global brand "ambassador" for UC was demonstrable, not only in RIM's devices, services and marketing campaigns, but in the economic data summarized in its annual report (RIM 2009). Indeed, 2009 was the tipping point for RIM and the BlackBerry as the smartphone market came to be increasingly dominated by Apple's iPhone and the various Android-based handsets. This tipping point, however, is indicative of the broader maturation of UC as a basic staple of everyday life, even in places (like on the African continent) where traditional consumer technologies had been ignored due to insufficient demand, high cost or lack of infrastructure (Arnquist 2009; Evans 2012; Wright 2008).

RIM's decline has fueled intense commentary since the launch of the iPhone. Rarely a day goes by without another obituary for RIM and the BlackBerry. The resignation of co-CEOs Mike Lazaridis and Jim Balsillie in 2012 further signaled the dramatic changes both in the company and its specific claim to a branded experience of UC. Whether these assessments are based on sound economic analysis, hysteria born from the "animal spirits" of a chaotic marketplace, or the need to provide regular content for business and technology blogs is debatable. Regardless of the reason, the signs of decline are palpable, though champions of the BlackBerry brand persist, and sales continue to grow in many developing markets (Africa, Latin America and Southeast Asia are still areas of growth for the BlackBerry). RIM, now rebranded as BlackBerry to create consistency between its product lines and corporate operations, attempted to reboot its brand of devices by launching an entirely new operating system and app development platform, BB10. The result was a spectacular failure and the future of the company and its devices is very much uncertain (cf. Silcoff et al. 2013).

Quite apart from the postmortems and prognostications concerning the company, our analytical point is that the question of RIM's decline is intimately tied its own role in the reification of UC as a now taken-for-granted expectation. Indeed, the success of the BlackBerry as a unique branded arguably has been occluded by its experiential universalization (at least in relatively "developed" political economies). Declining handset costs, increased processing power and expanded mobile bandwidth capacity all have enabled mobile devices to develop into ubiquitous platforms for the consumption of software and services. Thus ubiquitous connectivity is no longer a selling point for the company. Instead, the BlackBerry's success has contributed to its demise as its branded experience of ubiquitous connectivity has become embedded, and taken for granted, technical and experiential characteristic of mobile media generally. Though troubled, the company is returning to its roots by focusing on large-enterprise clients seeking secure and efficient mobile devices and services that optimize the collaborative and flexible aspects of the "virtual" organization. While the specific brands, devices and services may change, the essential features of BlackBerry™ Capitalism are now deeply rooted in the strategies of virtualization characteristic of contemporary capitalism.

BlackBerry™ Capitalism Is Dead. Long Live BlackBerry Capitalism!

NOTES

1. The source of the data for stock prices and market capitalization is http://ca.finance.yahoo.com/q/hp?s=BBRY&a=01&b=4&c=1999&d=01&e=27&f=2014&g=m&z=66&y=0andhttps://ycharts.com/companies/BBRY/chart/#/?securities=type:company,id:BBRY,include:true,,&calcs=id:market_cap,include:true,,&format=real&recessions=false&zoom=custom&startDate=1%2F1%2F2009&endDate=1%2F1%2F2014&chartView=respectively (both accessed February 27, 2014). The data for global market share in 2008 and 2014 came from www.forbes.com/sites/ewanspence/2013/01/30/blackberry-must-ignore-market-share/ and www.bnn.ca/News/2013/11/12/BlackBerrys-global-market-share-falls-to-just-17.aspx, respectively (both accessed February 27, 2014). The actual peak of RIM's valuation on the stock market was in August of 2007, when its stock price reached $236 per share with a market valuation of $69.3 billion. At its apex, the BlackBerry OS market share was 20% in 2009 (Silverman 2012)
2. In inserting the slash between "capital" and "ism" we want to give notice to the articulation between "capital" as an object and process, on the one hand, and the social structures of accumulation of capitalism as a mode of production that enable its production, distribution and consumption, on the other hand.
3. Some of the key texts in the literature on "digital capitalism" are McChesney (2004, 2008, 2013), Mosco (2005, 2009) and Schiller (1999, 2011).
4. See the introduction to this book for a more detailed discussion of materialist medium theory as well as Packer and Wiley (2011) and Parikka (2012).
5. One example would be the development of accountancy. For early-twentieth-century social theorists of capitalism such as Werner Sombart and Max Weber, the practice of double-entry bookkeeping (DEB) was fundamental to the establishment of the modern capitalist enterprise that manufactured, bought and sold different commodities across different times and places (Chiapello 2007). From the point of view of Innisian medium theory, DEB is not just an accounting practice but, more importantly, an articulation of elements both material (desks, pens, account ledgers, files, shelves, etc.) and imaginary (rules of mathematics, standards of accountancy, norms of professional accountants and actuaries, fiduciary ethics of corporate governance, etc.)—which comprise a social practice of mediated power-knowledge. Moreover, such practices give rise to cadres and elites, as Innis argued, which structure access to and control over such ensembles of power-knowledge and create monopolizations of knowledge (Innis 2007). As we shall see in the second part of this chapter, this applies as much to the Chief

Information Officers of the twenty-first-century corporate entities and the "crackberry" devotees of the BlackBerry they manage, as well as to the green-eye-shaded accountants of the nineteenth century.

6. We are by no means the first to suggest such a combination. The work on digital gaming by Nick Dyer-Witheford and Greig de Peuter (2009) exemplifies this articulation of theoretical traditions in important ways See also their earlier work with Stephen Kline (2003).

7. Some of the key texts of the informational capital/ism perspective are Arvidsson and Colleoni (2012), Fisher (2010), Fuchs (2008, 2010, 2011, 2012a, 2012b, 2013), Terranova (2004), Wittel (2012) and Wark (2012).

8. There is a growing literature on the concept of "immaterial labour," particularly in terms of how its central to the prosumerist productivity—and this corporate profitability—of social media platforms such as Facebook. For starters, see Andrejevic (2011, 2013), Cote and Pybus (2011), Fisher (2012), Peters and Bulut (2011), Scholz (2013), Terranova (2013) and an entire special issue of the journal *ephemera*, Burston, Dyer-Withford and Hearn (2010).

9. See the famous *BusinessWeek* cover story on "The Virtual Corporation" www.businessweek.com/stories/1993–02–07/the-virtual-corporation (accessed February 27, 2014). For later reflections (2013) on the "triumph" of the virtual corporation, see http://finance.yahoo.com/blogs/the-exchange/triumph-virtual-corporation-194905193.html (accessed Febraury 27, 2014). We will discuss this further in the next section.

10. A visual history of BlackBerry devices is available at http://crackberry.com.

11. By "colonization" we mean both the proliferation of available digital devices and services as well as their seamless embedding into the rhythms of everyday life.

12. Lexicon Branding also was responsible for coming up with Intel's "Pentium" brand as well as Apple's "PowerBook" line of laptop computers. See www.lexiconbranding.com.

13. As John Durham Peters (1999) has lucidly documented, the mythology of unmediated communication has a long history. Radical changes in communication technologies, for example the telegraph, are often linked to a perceived supernatural augmentation of human faculties. For example, Harvard physicist and Morse biographer John Trowbridge wrote in 1899, "Wireless telegraphy is the nearest approach to telepathy that has been vouchsafed ["revealed"] to our intelligence" (quoted in Peters 1999, 104).

14. Here we use the term "always on" in a double sense: the first one already articulated as a technological condition, the second I reference a more theatrical and performative sense of being *on* meaning that one is in effect always performing, whether it be in service of work, or in the iterative project of the self so essential to the consumerist ethos; see Bauman (2007).

15. On the scope of RIM's affective turn, McGuigan (2010) writes,

> Most striking about this campaign is its attempt to mobilize *affect*. The images characterize BlackBerry devices as an archive and conduit of 'everything we love in life.' Ironically, the images remind us that what we really value—what gives meaning to our lives—are interactions with the *people* we love and care about. Perhaps this is meant to tap into guilt experienced by people forced to spend time apart from their families—as is common in the corporate world. In this sense, the products and their connective capabilities are reified, serving as proxies or facilitators for interactions that cannot occur in physical proximity, for whatever reason(s). (20)

SOURCES

Andrejevic, M. "Social Network Exploitation." In *A Networked Self: Identity, Community, and Culture on Social Network Sites*, edited by Z. Papacharisi, 82–101. New York: Routledge. 2011.

Andrejevic, M. "Estranged Free Labor." In *Digital labor: The Internet as Playground and Factory*, edited by T. Scholz, 149–164. New York: Routledge, 2013.

Arminen, I. "Intensification of Time-Space Geography in the Mobile Era." In *The Reconstruction of Space and Time: Mobile Communication Practices*, edited by R. Ling and S. Campbell, 89–108. New Brunswick, NJ: Transaction Publishers, 2009.

Arnquist, S. "In Rural Africa, a Fertile Market for Mobile Phones." *The New York Times*. October 5, 2009.

Arvidsson, A. *Brands: Meaning and Value in Media Culture*. New York: Routledge. 2006.

Arvidsson, A., and E. Colleoni. "Value in Informational Capitalism and on the Internet." *The Information Society*. no. 3 (2012): 135–150.

Bailey, H. "POTUS Keeps His Precious." *Newsweek*. February 2, 2009. Accessed March 3, 2014. www.newsweek.com/president-obama-gets-keep-his-blackberry-77947

Baker, R. *Networking the Enterprise: How to Build Client/Server Systems that Work*. New York: McGraw-Hill, 1994.

Bauman, Z. *Consuming Life*. Cambridge: Polity, 2007.

Beniger, J. *The Control Revolution: Technological and Economic Origins of the Information Society*. Cambridge, MA: Harvard University Press, 1986.

Benkler, Y. *The Wealth of Networks: How Social Production Transforms Markets and Freedom*. Hartford, CT: Yale University Press, 2006.

Birchall, D. W., and L. Lyons. *Creating Tomorrow's Organization: Unlocking the Benefits of Future Work*. London: Pitman Pub, 1995.

Boltanski, L., and E. Chiapello. 2007. *The New Spirit of Capitalism*. New York: Verso, 2007.

Boutang, Y. *Cognitive Capitalism*. Cambridge, UK: Polity, 2011.

Brodsky, I. "Wireless Data Networks and the Mobile Workforce." *Telecommunications*. no. 24 (1990): 31.

Bruns. A. *Blogs, Wikipedia, Second Life, and Beyond From Production to Produsage*. New York: Peter Lang, 2008.

Burston, J, N. Dyer-Witheford and A. Hearn. "Digital Labour: Workers, Authors, Citizens." *ephemera: Theory and Politics in Organization*. no. 10 (2010). Accessed March 23, 2014. www.ephemerajournal.org/contribution/digital-labour-workers-authors-citizens

Byers, R. "The Ethical Implications of the Virtual Work Environment." In *Communication, Relationships and Practices in Virtual Work*, edited by Shawn D. Long, 68–86. Hershey, PA: Business Science Reference, 2010.

Carey, James. *Communication as Culture, Revised Edition: Essays on Media and Society*. New York: Routledge, 2008.

Castells, M. *The Rise of the Network Society*. Cambridge, MA: Blackwell Publishers, 1996.

Chakravartty, P., and D. Schiller. "Neoliberal Newspeak and Digital Capitalism in Crisis." *International Journal of Communication*. no. 4 (2010): 670–692.

Chiapello, E. "Accounting and the Birth of the Notion of Capitalism." *Critical Perspectives on Accounting*. no. 18 (2007): 263–296.

Clifford, S. "For BlackBerry, Obama's Devotion Is Priceless." *The New York Times*. January 9, 2009. Accessed March 13, 2014. www.nytimes.com/2009/01/09/business/media/09blackberry.html

Colapinto, J. "Famous Names: Does It Matter What a Product Is Called?" *The New Yorker.* October 3, 2011. Accessed September 1, 2104. http://www.newyorker.com/magazine/2011/10/03/famous-names

Comor, E. "Contextualizing and Critiquing the Fantastic Prosumer: Power, Alienation and Hegemony." *Critical Sociology.* no. 37 (2011): 209–327.

Conforti, J. "Somebody's Watching Me: Workplace Privacy Interests, Technology Surveillance, and the Ninth Circuit's Misapplication of the Ortega Test in Quon v. Arch Wireless." *Seton Hall Circuit Review.* no. 5 (2009): 1–36.

Cortada, J. *Information and the Modern Corporation.* Cambridge, MA: MIT Press, 2011.

Cote, M., and J. Pybus. "Learning to Immaterial Labour 2.0: Facebook and Social Networks." In *Cognitive Capitalism, Education, and Digital Labor,* edited by M. Peters and E. Bulut, 169–194. New York: Peter Lang, 2011.

Davidow, W., and M. Malone. *The Virtual Corporation: Structuring and Revitalizing the Corporation for the 21st Century.* New York: Edward Burlingame Books/HarperBusiness, 1992.

Dewar, T. "Virtual Teams—Virtually Impossible?" *Performance Improvement.* no. 45 (2006): 22–25.

Didner, B. "Data Goes Mobile." *Telephony.* no. 22 (1991): 24–25.

Dyer-Witheford, N. 1999. *Cyber-Marx: Cycles and Circuits of Struggle in High Technology Capitalism.* Urbana-Champaign, IL: University of Illinois Press, 1999.

Dyer-Witheford, N., and G. de Peuter. *Games of Empire: Global Capitalism and Video Games.* Minneapolis: University of Minnesota Press, 2009.

Easton, J. *Going Wireless: Transform Your Business with Mobile Technology.* New York: HarperBusiness, 2002.

Ebrahim, N., S. Ahmed and Z. Taha. "Virtual R&D Teams in Small and Medium Enterprises: A Literature Review." *Scientific Research and Essays.* no. 4 (2009): 1575–1590.

Edwards, M. "Wireless Internet Isn't Here Yet—But it's Coming." *Communications News.* no. 35. June (1998): 3.

Elkin, T. "BlackBerry: Mark Guibert." *Advertising Age.* no. 72 (2001): 28.

Evans, J. "In Five Years, Most Africans Will Have Smartphones." *TechCrunch.* June 9, 2012. Accessed July 10, 2012. http://techcrunch.com/2012/06/09/feature-phones-are-not-the-future/

Faigen, G., B. Fridman and A. Emmett. *Wireless Data for the Enterprise: Making Sense of Wireless Business.* New York: McGraw-Hill, 2002.

Fisher, E. *Media and New Capitalism in the Digital Age.* London: Palgrave Macmillian, 2010.

Fisher, E. "How Less Alienation Creates More Exploitation? Audience Labour on Social Network Sites," *tripleC: Open Access Journal for a Global Sustainable Information Society.* no. 10 (2012): 171–183.

Fitzpatrick, T. "Critical Theory, Information Society and Surveillance Technologies." *Information, Communication & Society.* no. 5 (2002): 357–378.

Fortunati, L. "The Mobile Phone as Technological Artefact." In *Thumb Culture: The Meaning of Mobile Phones for Society,* edited by P. Goltz, S. Bertschi and C. Locke, 149–160. Piscataway, NJ: Transcript, 2005.

Foucault, M. *Power/Knowledge: Selected Interviews and Other Writings, 1972–1977.* New York: Vintage, 1980.

Fukuyama, F., A. Shulsky, United States Army, Arroyo Center, and Rand Corporation. *The "Virtual Corporation" and Army Organization.* Santa Monica, CA: Rand, 1997.

Fuchs, C. *Internet and Society: Social Theory in the Information Age.* London, UK: Routledge, 2008.

Fuchs, C. "Labour in Informational Capitalism." *The Information Society*. no. 26 (2010): 176–196.

Fuchs, C. *Foundations of Critical Media and Information Studies*. London, UK: Routledge, 2011.

Fuchs, C. "Dallas Smythe Today: The Audience Commodity, the Digital Labour Debate, Marxist Political Economy, and Critical Theory." *tripleC: Cognition, Communication, Co-Operation*. no. 10 (2012a): 692–740.

Fuchs, C. "With or without Marx? With or without Capitalism? A Rejoinder to Adam Arvidsson and Eleanor Colleoni." *tripleC: Cognition, Communication, Co-operation*. no. 10 (2012b): 633–645.

Fuchs, C. *Social Media: A Critical Introduction*. Los Angeles: Sage, 2013.

Fuchs, C., and S. Sevignani. "What Is Digital Labour? What Is Digital Work? What's Their Difference? And why do these questions matter for Understanding Social Media?." *tripleC: Cognition, Communication, Co-operation*. no. 11 (2013): 237–293.

Glotz, P., S. Bertschi and C. Locke. *Thumb Culture: The Meaning of Mobile Phones for Society*. Bielefeld: Transcript, 2005.

Goggin, G. *Cell Phone Culture: Mobile Technology in Everyday Life*. New York: Routledge, 2006.

Gowdy, J., and M. Walton. "Consumer Sovereignty, Economic Efficiency and the Trade Liberalisation Debate." *International Journal of Global Environmental Issues*. no. 3 (2013): 1–13.

Grenier, R., and G. Metes. *Going Virtual: Moving Your Organization Into the 21st Century*. Upper Saddle River, NJ: Prentice Hall, 1995.

Haigh, T. "Technology, Information and Power: Managerial Technicians in Corporate America, 1917–2000." PhD diss., University of Pennsylvania, Philadelphia, 2003.

Hamblen, M. "The BlackBerry: In Search of the 'Prosumer.'" *Computerworld*. May 14, 2008. Accessed September 1, 2014. http://www.computerworld.com/article/2535877/mobile-wireless/the-blackberry--in-search-of-the--prosumer-.html

Harmon, Amy. "E-mail You Can't Outrun." *The New York Times*. September 21, 2000.

Harvey, D. *The Condition of Postmodernity: An Enquiry into the Origins of Cultural Change*. Malden, MA: Blackwell, 1990.

Harvey, D. *The Enigma of Capital and the Crises of Capitalism*. New York: Oxford University Press, 2010.

Headrick, D. *When Information Came of Age: Technologies of Knowledge in the Age of Reason and Revolution, 1700–1850*. New York: Oxford University Press, 2002.

Healy, B. "Entrepreneurs Grab Latest Handheld Technology." *The Globe and Mail*. July 13, 2000. Accessed September 1, 2014. http://www.theglobeandmail.com/incoming/entrepreneurs-grab-latest-handheld-technology/article1040958

Hengel, G. "Cellular Communication Key to Mobile Workforce." *National Underwriter*. November 7, 1994.

Herman, A. "Production, Consumption and Labor in the Social Media Mode of Production." In *The Social Media Handbook*, edited by J. Hunsinger and T. Senft, 30–44. New York: Routledge, 2013.

Hesselbein, F., M. Goldsmith, R. Beckhard and Peter F. Drucker Foundation for Nonprofit Management. *The Organization of the Future*. San Francisco: Jossey-Bass, 1997.

Huws, U., W. Korte and S. Robinson. *Telework: Towards the Elusive Office*. Toronto: Wiley, 1990.

Innis, H. *Empire and Communications*. Toronto: Dundurn Press, 2007.

Innis, H. *The Bias of Communication*, 2nd ed. Toronto: University of Toronto Press, 2008.

Jackson, P. *Virtual Working: Social and Organisational Dynamics.* London: Taylor & Francis, 1999.

Jackson, P., and J. Wielen. *Teleworking: International Perspectives: From Telecommuting to the Virtual Organisation.* London; New York: Routledge, 1998.

Kline, R. "Cybernetics, Management Science, and Technology Policy: The Emergence of 'Information Technology' as a Keyword, 1948–1985." *Technology and Culture.* no. 47 (2006): 513–535.

Kline, S., N. Dyer-Witheford and G. de Peuter. *Digital Play: The Interaction of Technology, Culture, and Marketing.* Montreal: McGill-Queen's University Press, 2003.

Lazaridis, M. [Mike Lazaridis oral history]. *Computerworld Honors Program International Archives.* April 30, 2008. Accessed March 10, 2014. www.cwhonors.org/archives/histories/lazaridis.pdf

Lyon, D. *The Electronic Eye: The Rise of Surveillance Society.* Minneapolis: University of Minnesota Press, 1994.

MacNamara, W. "Why the BlackBerry Wasn't a Strawberry." *The Times.* June 6, 2012. Accessed March 10, 2014. www.thetimes.co.uk/tto/business/industries/supportservices/article3436690.ece

Mager, A. "Algorithmic Ideology: How Capitalist Society Shapes Search Engines." *Information, Communication and Society.* no. 15 (2012): 769–787.

Maney, K. "BlackBerry: The Heroin of Mobile Computing." *USA Today.* May 7, 2001.

Manzerolle, V. *Brave New Wireless World.* PhD diss. University of Western Ontario, London, Ontario, 2013.

Marx, K., and B. Fowkes. *Capital: A Critique of Political Economy.* Harmondsworth, UK: Penguin, 1976.

Mayr, O. "Adam Smith and the Concept of the Feedback System: Economic Thought and Technology in 18th-Century Britain." *Technology and Culture.* no. 12 (1971): 1–22.

McChesney, R. *The Problem of the Media: U.S. Communication Politics in the Twenty-First Century.* New York: Monthly Review Press, 2004.

McChesney, R. *The Political Economy of Media: Enduring Issues, Emerging Dilemmas.* New York: Monthly Review Press, 2008.

McChesney, R. *Digital Disconnect: How Capitalism is Turning the Internet Against Democracy.* New York: New Press, 2013.

McGuigan, L. "Ubiquitous and Loving It." Unpublished paper. University of Western Ontario. London, Ontario, 2010.

McLuhan, M. *Understanding Media: The Extensions of Man.* New York: McGraw-Hill, 1964.

Metselaar, C., and R. van Dael. "Organisations Going Virtual." *AI & Society.* no. 13 (1999): 200–209.

Michaluk, K., M. Trautschold, G. Mazo, J. Markham, C. Andres and S. Anglin. *Crackberry: True Tales of BlackBerry Use and Abuse.* New York: SpringerLink, 2011.

Micklethwaith, J., and A. Wooldridge. *The Witch Doctors: Making Sense of the Management Gurus.* New York: Times Books, 1996.

Morgan, G. *Imaginization: The Art of Creative Management.* Newbury Park, CA: Sage Publications, 1993.

Mosco, V. *The Digital Sublime: Myth, Power, and Cyberspace.* Cambridge, MA: MIT Press, 2005.

Mosco, V. *The Political Economy of Communication,* 2nd ed. Los Angeles: Sage, 2009.

Mowshowitz, A. "Virtual Organization: A Vision of Management in the Information Age." *The Information Society.* no. 10 (1994): 267–288.

Newitz, A. "The Boss Is Watching Your Every Click." *New Scientist*. no. 191 (2006): 30–31.

Nye, D. *American Technological Sublime*. Cambridge, MA: MIT Press, 1996.

Packer, J., and S. Crofts Wiley. *Communication Matters: Materialist Approaches to Media, Mobility and Networks*. London: Routledge, 2011.

Parikka, J. "New Materialism as Media Theory: Medianatures and Dirty Matter." *Communication and Critical/Cultural Studies*. no. 9 (2012): 95–100.

Pariser, E. *The Filter Bubble: What the Internet Is Hiding From You*. London, New York: Viking, 2011.

Peck, J. *Constructions of Neoliberal Reason*. New York: Oxford University Press, 2012

Peters, J. D. *Speaking into the Air: A History of the Idea of Communication*. Chicago: University of Chicago Press, 1999.

Peters, M., and E. Bulut. *Cognitive Capitalism, Education, and Digital Labor*. New York: Peter Lang, 2011.

Pratt, J. "Counting the New Mobile Workforce." Washington, DC: Department of Transportation, Bureau of Transportation Statistics, 1997.

Quinn, J. *Intelligent Enterprise: A Knowledge and Service Based Paradigm for Industry*. Toronto: Free Press, 1992.

Reeves, R. "Constant Connectivity in a Wireless Age: The Discursive Promotional Strategies of the Blackberry." Masters of Arts (Media Studies) thesis. Concordia University, Montreal, Quebec, Canada, 2007.

RIM. *Annual Report 1998*. Waterloo: Research in Motion, 1999.

RIM. *Annual Report 2000*. Waterloo: Research in Motion, 2001.

RIM. *Annual Report 2001*. Waterloo: Research in Motion, 2002.

RIM. *Annual Report 2003*. Waterloo: Research in Motion, 2004.

RIM. *Annual Report 2005*. Waterloo: Research in Motion, 2006.

RIM. *Annual Report 2006*. Waterloo: Research in Motion, 2007.

RIM. *Annual Report 2008*. Waterloo: Research in Motion, 2009.

RIM. *Annual Report 2009*. Waterloo: Research in Motion, 2010

Ryan, A. "The Whole World in Your Hand?" *Computerworld*. no. 25 (1991): 65.

Sbihli, S. *Developing a Successful Wireless Enterprise Strategy a Manager's Guide*. New York: Wiley, 2002.

Schiller, D. *Digital Capitalism: Networking the Global Market System*. Cambridge, MA: MIT Press, 1999.

Schiller, D. *How to Think about Information*. Urbana-Chamapign: University of Illinois Press, 2006.

Schiller, D. "Power Under Pressure: Digital Capitalism In Crisis." *International Journal of Communication*. no. 5 (2011): 924–941.

Scholz, T. *Digital Labor: The Internet as Playground and Factory*. New York: Routledge, 2013.

Schumpeter, J. *Capitalism. Socialism and Democracy*, 3rd ed. New York: HarperCollins, 2008.

Silcoff, S., J. McNish, and S. Ladurantaye. "Inside the Fall of BlackBerry: How the Smartphone Inventor Failed to Adapt." *The Globe and Mail*, September 27, 2013. Accessed Septemeber 1, 2014. http://www.theglobeandmail.com/report-on-business/the-inside-story-of-why-blackberry-is-failing/article14563602/#dashboard/follows/

Silverman, M. "The Rise and Fall of RIM." *Mashable*, June 25, 2012. Accessed September 1, 2014. http://mashable.com/2012/06/25/rim-decline-chart/

Star Staff. "Barack Obama Still a BlackBerry User." *Toronto Star*. December 5, 2013. Accessed March 3, 2014. www.thestar.com/business/2013/12/05/barack_obama_still_a_blackberry_user.html

Steinberg, M. *The Fiction of a Thinkable World: Body, Meaning, and the Culture of Capitalism*. New York: Monthly Review Press, 2005.

Sweeny, A. *BlackBerry Planet: The Story of Research in Motion and the Little Device that Took the World by Storm*. Hoboken, NJ: Wiley, 2009.

Tapscott, D. *The Digital Economy: Promise and Peril in the Age of Networked Intelligence*. New York: McGraw-Hill, 1996.

Terranova, T. *Network Culture: Politics for the Information Age*. London: Pluto Press, 2004.

Terranova, T. "Free Labor." In *Digital Labor: The Internet as Playground and Factory*, edited by T. Scholz, 33–57. New York: Routledge, 2013.

Toffler, A. *The Third Wave*. New York: Morrow, 1980.

Toffler, A. *The Third Wave: The Classic Study of Tomorrow*. 2nd ed. New York: Bantam, 1984.

Tomlinson, J. *The Culture of Speed: The Coming of Immediacy*. Los Angeles: Sage, 2007.

Turow, J. *The Daily You: How the New Advertising Industry Is Defining your Identity and Your Worth*. New Haven: Yale University Press, 2011.

Verespej, M. "The Anytime, Anyplace Workplace." *Industry Week*. no 13 (1994): 243.

Wark, M. *Telesthesia: Communication, Culture and Class*. Malden, MA: Polity, 2012.

Wasserman, T. "The Fruits of His Labor." *Brandweek*. no. 42 (2001): 44–48.

Watson M., N. Sridhar and H. Rhee. "Communication and Coordination in the Virtual Office." *Journal of Management Information Systems*. no. 14 (1998): 7–28.

Watson-Manheim, M., K. Chudoba and K. Crowston. "Discontinuities and Continuities: A New Way to Understand Virtual Work." *Information Technology & People*. no. 15 (2002): 191–209.

Webster, F. *Theories of the Information Society*, 3rd ed. New York: Routledge, 2006.

Wheeler, W. *Integrating Wireless Technology in the Enterprise*. Boston, Oxford: Elsevier/Digital Press, 2004.

Wintrob, S. "Wireless Firms Eye Growing Small-Business Market." *The Globe and Mail*. April 27, 2001. Last accessed May 5, 2001. www.theglobeandmail.com/incoming/wireless-firms-eye-growing-small-business-market/article1031311/#dashboard/follows/

Wittel, A. "Digital Marx: Toward a Political Economy of Distributed Media." *Triple C: Cognition, Communication, Co-operation*. no. 10 (2012): 313–333.

Wright, T. "Poorer Nations Go Online on Cellphones." *Wall Street Journal*. December 5, 2008.

7 Mobile Web 2.0
New Imaginaries of Mobile Internet

Gerard Goggin

We are witnessing an era of intense reliance on communication and media technologies as a central feature of social life. While debate rages about the extent, nature and implications of this development, it is something that not only is observable in private and public spheres but also is a preoccupation in a range of settings. Two of the most important technologies involved in these transformations are the Internet and mobile phones. Launched in 1969, the Internet developed steadily through the 1970s and 1980s before its widespread adoption from the early 1990s onward. The mobile phone had its roots in wireless telegraphy, radio and the telephone, was made available commercially in the late 1970s, and, with the second generation standard (2G) was adopted globally also during the 1990s. Both the Internet and mobiles are now established as mature global media technologies, underpinning the disruption of other communication forms and older media. What is remarkable about these twin trajectories is that they are now entwined as "mobile Internet"—the subject of this volume. Among the highly significant implications of this melding of mobiles and Internet is that mobiles now constitute a preferred way for many people in the world—especially the poor and those on low incomes—to access Internet.

There are many questions raised by the emergence and rise of mobile Internet, which is at the center of cultural and social transformations in many countries. My focus in this chapter is on the distinctive and novel manner in which the social is being re-created (or cocreated) along with the production, consumption and shaping of these new technologies. To understand this, I am interested in considering how mobile Internet as an ensemble of technologies is imagined, and how it participates in wider social imaginaries.

Imagination is a potent, evocative concept in understanding human society and history. The recognization of imagination has often guided studies of new media (Balsamo 2011; Boddy 2004; Douglas 2004; Kirkpatrick 2013; Wheeler 2006). Imagination is closely related to the more delimited concept of "imaginary." Used across a range of philosophical, social and cultural inquiries (Wilson and Dissanayake 1996; Le Doeuf 1989; Taylor 2004), an "imaginary" refers to a cluster of interrelated ideas. While not necessarily an "ideology" or possessing an ideological function, an "imaginary" is a pervasive, widely held attitude, perception or set of beliefs that colors the way something is regarded, understood or even shaped. Paradoxically, perhaps, in this sense an "imaginary" is not just "imaginary" (that is, conforming to the dictionary definition of only existing in the imagination). Rather, an imaginary has a material existence, as well as real influence and force.

In this chapter, I will be especially drawing upon the philosopher Charles Taylor's concept of "social imaginary" (Taylor 2004). Taylor notes that an imaginary is a contradictory, ambiguous thing. It has a relationship to actual practices, and also new possibilities. It also is the matrix of values and norms that frame an understanding of the world. For Taylor, the "social imaginary is not a set of ideas; rather, it is what enables, through making sense of, the practices of a society" (Taylor 2004, 2). Taylor suggests that social imaginaries:

> have a constitutive function, that of making possible the practices that they make sense of and thus enable . . . Like all forms of human imagination, the social imaginary can be full of self-serving fiction and suppression, but it also is an essential constituent of the real.
>
> (Taylor 2004, 184)

Taylor argues that Western modernity is characterized by a "new conception of the moral order of society," which mutates into a social imaginary spawning our characteristic social forms such as the "market economy, the public sphere, and the self-governing people, among others" (Taylor 2004, 2). One of Taylor's key insights is the importance of social imaginaries for helping people make sense of transformations. Discussing the rise of modern individualism as "by its very essence a solvent of community," he notes that with the French Revolution we can see people "expelled from their old forms—through war, revolution, or rapid economic change—before they can find their feet in the new structures, that is, connect some transformed practices to the new principles to form a viable social imaginary" (Taylor 2004, 18).

Taylor's account of social imaginaries is especially suggestive for our topic of mobile Internet. In mobile phones, we have a flexible, versatile, powerful and very widely used set of technologies that are not only tools for negotiating everyday life for billions of people—they support practices and meanings about people's lives, and where they fit into structures of power. Unsurprisingly, there is an extensive literature on the social and cultural significance of mobile phones, and what they represent. A new development in mobiles since at least 2007 onward has been the pervasive of mobile Internet technologies, which are now highly important to how "ordinary people 'imagine' their social surroundings," as Taylor puts it (Taylor 2004, 23). My argument is there is a new social imaginary emerging, associated with and centring upon mobile Internet—correlated to a set of moral norms and values for society in general—a "common understanding that makes possible common practices and a widely shared sense of legitimacy" (Taylor 2004, 23). Mobile Internet suggests a new category, which subtends the constitutive tensions in late modernity that Taylor identifies in his account of modern social imaginaries. In particular, we find a tension between the notion of sharing, central to contemporary social media, and mobile Internet media, on the one hand, and the logics of economic individualism and social exclusion that underpin this period of digital technologies. Crucially mobile Internet takes shape differently in particular national and cultural contexts, and modernities—just as Taylor notes that "modern social imaginaries have been differently refracted in the divergent media of the respective national histories" (Taylor 2004, 154). Indeed the social imaginaries in which mobile Internet are implicated are especially interested because they engage a range of modernities—Western, non-Western, and others.

MOBILE INTERNET ASSEMBLAGES

In itself, mobile Internet is a complex assemblage (Bennett and Healy 2011)—a set of media and communication ecologies that has, until recently, slowly taken shape. With the popularity of the Web and its characteristics—ease of use, linking the resources of the Internet, working across devices, platforms, applications and screen—quickly established by the mid-1990s, the mobile phone was an obvious platform for the industry to concentrate its efforts (Nokia, 2000). Finnish giant Nokia, along with Motorola, Ericsson and Unwired Planet (later to become Phone.com), instigated the Wireless Application Protocol (WAP) forum in 1997 (Kumar, Parimi and Agrawal 2003; WAP Forum 2000). WAP was a key way that mobile manufacturers and carriers sought to implement an Internet-like environment on mobile phones. Rather than the mobile phone functioning like any other Internet-connected device (for instance, the way Wi-Fi operated), WAP extended a Web client to the device—a crucial difference from the Internet TCP/IP underpinning the public Internet (Huston 2001). In any case, due to slow data speeds and limitations on handsets, operating systems, and applications, WAP proved a frustrating experience for users—and only attracted limited interest (Goggin 2006; Helyar 2001).

The celebrated pioneer in mobile Internet, however, was the i-Mode "ecosystem," developed by the Japanese carrier NTT DoCoMo (Natsuno 2003). The i-Mode system was a packet-switched data service that operated over the mobile phone network. The innovation in i-Mode lay in the creation of close relationships between networks (owned by DoCoMo) and the other crucial elements of the whole mobile data environment, such as handsets, gateways, servers (with the billing systems), portals and content. Content providers were encouraged to develop products (such as mobile music), and purchase of the service by consumers was made as easy as possible—a notable achievement in the early days of mobile Internet. It took some years before the same ease of use was available in other country—when WAP did become widely used, in its incarnation as WAP 2.0—as mobile portal and premium services took off in 2002–2005, with music, video and other downloads, and multimedia and text messaging proved mainstays of mobile services (Spurgeon and Goggin 2007). This eventual take-up of WAP, harnessed to different business models (Ramos et al. 2002), was one reason why i-Mode did not prove exportable in its original form (Maitland, Bauer and Westerveld 2002; Goggin 2006)—although the approach, applications and services which it incubated are obvious precursors to Apple's iPhone, Google's Android, and other smartphones and apps platforms (Goggin 2011b).

It is probably more accurate to see the history of mobile Internet as an evolutionary process, but it is certainly possible to discern two periods of intense development and activity. So far I have discussed the period of the late 1990s and early 2000s, where there was focused, coordinated activity on conceiving, designing and implementing mobile Internet. The second period dates from roughly the middle of the first decade of the 2000s, we can point to an overlapping set of developments that less than five years later made mobile Internet a very widely discussed focus for directions in digital technology—and which greatly shifted the social imaginary of mobile Internet.

To start with, networks developed greatly, with cumulative effects. From a difficult start, third generation networks (3G) were widely rolled out, supporting higher data transfer rates. Existing 2G networks were also extended to better support mobile data. Thus Internet could be much more easily accessed via mobile networks—including mobile video, games, music, photo-sharing sites and the other facets of contemporary "broadband" Internet experience (in the Global North, at least). Mobile networks were also used to provide mobile broadband for laptop computers, tablets and other devices. That is, a chip or USB modem was used in conjunction with a computing device to provide broadband Internet. Mobile broadband achieved very rapid take-up around the world. Fourth generation (4G) mobile networks—involving a mix of mobile cellular and wireless Internet (Wi-Max, and other successor technologies for Wi-Fi) technologies—promised to make much faster access Internet speeds a reality. Finally, developments in next-generation broadband networks accelerated the process of replacing traditional fixed and mobile telecommunications circuits with Internet protocol based packet-switch networks—with significant implications for mobile Internet also (Middleton and Given 2011).

Another important factor in the second phase of mobile Internet is the continued development of mobile handsets. Multimedia handsets predominated in markets where consumers could afford them. With the success of Apple's iPhone, launched in mid-2007, and development of Google's Android operating system, smartphones became very fashionable. Smartphones were especially predicated on mobile Internet access, allowing downloading and use of "apps." Indeed the whole area of applications for mobiles gained dramatic impetus. As well the phenomenon of smartphones and other multimedia phones providing a much better platform for applications, the advent of Internet-based "social media" engaged mobiles in particular.

In Asian countries, mobiles have long been important in Internet access and use—especially in their pioneering of social software, with long-established communities around applications such as Cyworld (South Korea) and Mixi (Japan) (Hjorth 2009). In the West, social networking systems had been developed around desktop Internet platforms until comparatively recently (Boyd and Ellison 2007)—with exceptions such as mobile social software, experimented with from the late 1990s (Humphreys 2008). So it was not until 2007–2008 that social networking systems, software and social media become widespread on mobiles in non-Asian countries—and then the growth was phenomenal. By 2010, Facebook had established itself as the leading social networking system in the West, with a substantial proportion of users accessing it on mobiles. In the process, it also become a platform adopted, and reshaped, by users in non-Western countries—something underscored by its role in the "Arab Spring" uprisings of late 2010 and early 2011, which were claimed to be led by "Facebook revolutionaries"—the "Facebook generation" that transcended "classical" political movements (Beaumont and Sherwood 2011; Shenker 2011). The prominence of Facebook's rise, of course, obscured the fact that there were a myriad of other social networking and social media applications used across the world, not least in markets like China (Qiu 2009; Yu 2009).

Finally, the second phase of mobile Internet was characterized by the interaction of a wide range of different networks and devices—over and above mobiles and Internet. This is evident with the rise of locative media, based on positioning, locational and mapping technology, such as Global Positioning Satellites (GPS), Google Maps and Earth (and other "geospatial Web") (de Souza e Silva and Frith, 2012; Farman, 2012; Gordon and de Silva e Souza, 2011). Also fast emerging were networks of sensing technologies, Radio Frequency ID (the long-awaited "Internet of Things"), and other technologies. Mobile Internet, then, becomes an especially complex assemblage in its second phase, from 2005 onward (Goggin 2009). It is constituted from a series of quite contingent interactions between different kinds of networks, devices, applications and practices. Mobile television, for instance, becomes as much about the possibilities of using a mobile phone to record video and then upload it to YouTube, as it does about broadcasting television programs to a user's mobile phone (which was the mobile industry's early vision). Or mobile Internet is as much about someone using an app to collect the statistics on their bicycle journey to work ("map my ride"), and sharing that with others, as it is about using their phone to browse the Web.

Mobile Internet is also very much bound up, from 2012 onward, with the rise of the pervasive data creation, collection, processing and harvesting of what is commonly—and obviously problematically—referred to as "big data." The affordances of mobile Internet for such nigh-compulsory everyday data surveillance were dramatized in 2013–2014, through the many relevations of the leaks of material made publicly available from U.S. whistle-blower Edward Snowden. Take, for instance, one exemplary allegation that the U.S. National Security Agency (NSA) and its British counterpart GCHQ were developing capabilities to harvest data from "leaky" smartphone apps, such as the famous *Angry Birds* game:

> The data pouring onto communication networks from the new genera-tion of iPhone and Android apps ranges from phone model and screen size to personal details such as age, gender and location. Some apps, the documents state, can share users' most sensitive information such as sexual orientation—and one app recorded in the material even sends specific sexual preferences such as whether or not the user may be a swinger . . . Scooping up information the apps are sending about their users allows the agencies to collect large quantities of mobile phone data from their existing mass surveillance tools—such as cable taps, or from international mobile networks—rather than solely from hacking into individual mobile handsets.
>
> (Ball 2014)

As the work of many scholars makes clear—especially the pioneering work of Mark Andrejevic (2007, 2013)—such "sharing" of personal infor-mation is the well-entrenched dark side of the affordance of social and mobile media, a topic to which I will shortly return.

IMAGINING MOBILE INTERNET

While mobile phones have been the subject of great promises and hopes for sometime—ending poverty, for instance, as Iqbal Quadir, banker and founder of Grameen Phone, famously suggested in his widely noticed 2007 TED talk (Quadir 2005; cf. Toyama 2010)—it is fair to say that the Internet has been figured as a sublime medium, since the invention of the World Wide Web and its mass diffusion from 1991 onward. This is evident in one of the most detailed studies of the development of ideas about the Internet, Patrice Flichy's *The Internet Imaginaire* (Flichy 2007). Flichy sets out to chart the emergence of a set of ideas associated with, and influencing the development of the Internet—the discourses that form an "integral part of the develop-ment of a technical system" (Flichy 2007, 2). Because of the significance of the U.S. to how the Internet emerges in its early phase, he focuses upon the American context. As he notes, the imaginary he documents is quite specific to the American context—while influential elsewhere. Indeed Flichy remarks how the Internet *imaginaire* was "born in the particular context of the United States but subsequently became universal" (Flichy 2007, 211). In particular Flichy proposes that there is a "cyber-*imaginaire*," produced from the "technological *imaginaire*" associated with the technical projects of conceiving and developing the Internet (Flichy 2007, 107). Key ideas in the composition of this cyber-*imaginaire* can be traced back an elite of thinkers on the future initially (the likes of Marshall McLuhan, George Gilder, Nicholas Negroponte and others), but once it achieves cohesion and force, it operates much more broadly as the dominant way to imagine the Internet.

If we can transpose the linguistically and culturally specific concept of the *imaginaire*, and Flichy's work on the emergence of the Internet, to our contemporary terrain of the mobile Internet some two decades on, then we can find both interesting resonances and divergences. In the first phase of mobile Internet in the late 1990s to early 2000s, when the idea of such technology is not widely appreciated, we find a technological imaginary being shaped. Hence Nokia's proposal:

> We are witnessing the transformation of the mobile phone from a voice-centric communication device to a tool for managing business and private life, and for triggering and sharing experiences . . . In our vision, the Internet will go mobile, just as voice communication has done.
>
> (Nokia 2000, 2)

The larger social imaginary associated with early mobile Internet is predominantly that of the information society (something explicitly noted in Nokia's 2000 *WAP White Paper*). The information society looms large in the pioneering Finnish social theory of mobiles (Kopomaa 2000). The information society also proves a serviceable and durable concept for orienting the larger European intellectual and policy frameworks in which mobile Internet developments are being grasped, as part of greater ensembles of technological systems and notions of the social.

Interestingly, of course, the information society as a concept was greatly influenced by Japanese efforts to grasp and theorize the interrelationships between technology and society in the 1960s (Webster 2002). Yet when it comes to the Japanese i-Mode, we find a distinctive tone to the celebration of this pioneering technology and the culture that produced it (Barnes and Huff 2003; Funk 2001)—often partaking of what David Morley and Kevin Robins term a "techno-orientalism" (Morley and Robins 1995). Writing on the vogue for i-Mode, Mizuko Ito situates it in a deep-rooted "Euro-American fascination with Japanese technoculture," that invokes Japan as an "alternative technologized modernity":

> On the one hand, i-mode is held up as a technological and business model to be emulated; on the other hand, discourse abounds on the cultural strangeness of Japanese technofetishism that casts it as irreducibly foreign.
>
> (Ito 2005, 2)

Figure 7.1   "Mobile Internet with the Best Coverage . . . The Internet Where All Is Possible" (Claro advertising billboard, Cusco, Peru, 2011).

It is worth bearing in mind the specific nature of these social imaginaries attendant on the first phase of mobile Internet—European and Japanese—as we now move to consider second, current stage of mobile Internet, and the social imaginaries associated with it.

Let me provide two texts that illustrate key aspects of this social imaginary of contemporary mobile Internet, both drawn from advertising in Latin America. The first text is an advertisement for mobile Internet (Figure 7.1), offered by Claro, an interest of the Mexican giant América Móvil, owned by the world's richest man, Carlos Slim Helú (Goggin 2011a, 23). Mobile Internet, especially mobile broadband, is highly significant in Latin America, where it is the preferred type of Internet connection (Flores-Roux and Mariscal 2011, 5; Mariscal, Gamboa and Rentería Marín 2014).

The ad targets travelers to the iconic Incan city of Machu Picchu, at the time being celebrated for its one-hundredth-year anniversary of its "scientific discovery" by Hiram Bingham, the North American explorer and Yale University professor (Bingham 1948). A young woman, superimposed on one of the high views of Machu Picchu (perhaps from the Sun Gate, through which hikers on the so-called Inca Trail arrive at the site), has a double purchase on its splendors . She enjoys contemplating Machu Picchu with her own eyes, with a version of the image also displayed on the screen of her laptop. What is especially striking and amusing about this advertisement is how awkward and cumbersome the technology of mobile Internet seems here (indeed an ironic counterpoint to the subtle and powerful technologies the Incan society produced, in testaments to their culture such as Machu Picchu). The conceit of the backpacker sees the USB modem carried on her back, but expanded to the full size of a backpacker. She carries her laptop as commonly held, cradled in her arms, but, again, achieving connectivity and coverage at a fair price to traveler ease and comfort. The motto of the advertisement sings the praises of the "Internet where all is possible" ("La red donde todo es posible"), but the grounds of this utopia lie firmly in the crushing immobility of lugging around the burdensome combination still needed of laptop and mobile broadband device.

Figure 7.2   "With an Android Tigo, You're in Everything" (Tigo advertising billboard, Cartegna, Columbia, 2011).

If the backpacker is a suitable emblem for the harnessing of mobiles and Internet, with mobile broadband, advertisements for smartphones typically offer a different take. An advertisement for a Motorola smartphone using Google's Android operating system emphasizes that with such a device "you're in everything" (Figure 7.2).

This text nicely underscores the claims and potential experiences of smartphones by recoding them into categories redolent of the immersive, interactive viewing and audience modes and experience of film and television. The action adventure on the small screen literally spills over the frame of the phone, rather like the trope of the television program erupting into the viewer's lounge room. The "Weather Channel" icon reminds us that the smartphone provides recognizable television programming and content as well as Internet and new kinds of apps. For many, the mobile phone is no longer an adjunct, but indeed an essential tool and cherished object for navigating the rapids of society.

These two advertising texts provide examples of two prominent representations of mobile Internet, with strong continuity with preceding practices and ideas of the Internet (as evident with the modem-equipped backpacker at Machu Picchu) and mobile devices (the white-water rafting of the Android). Both also suggest new possibilities for mobile Internet. The first image struggles to resolve the tensions between the unwieldiness of the technology, on the one hand, and the infinite possibilities it affords, on the other hand. The second image draws on the practices and representations of mobile phones, established now over three decades. It also draw upon the long histories of media spectatorship and engagement, and the codes and repertoires by which they have found their way into advertising, to provide a new twist—on the small screen on the smartphone. Both these texts provide us with good examples of constituent elements of the emergent social imaginary emergent associated with mobile Internet. However, there is another category that, in my mind, really animates and distinguishes this imaginary that increasingly we find as a feature of discourse on mobile Internet: the concept of sharing.

THE POLITICS OF SHARING

At the heart of this new social imaginary is sharing—and how we understand this. In relation to mobiles, we see this imaginary emerging from the late 1990s, especially in relation to practices of text messaging, music, and photo and image sharing, and being capitalized upon especially in the smartphone era. Hence the title of this essay taken from a Peruvian billboard advertising smartphones offered by the Spanish mobile giant Movistar. (Movistar is owned by the parent company Telefónica, which dominates the market in Spain and has a strong presence in over a dozen countries in Latin America [Martínez 2008]). The advertisement (Figure 7.3) depicts a young man showing a young woman something on his phone, both close together and smiling, with the tagline "shared, life is more."

Figure 7.3   "Shared, Life is More" (Movistar advertising billboard, Lima, Peru, 2011).

The theme of sharing is something that recurs in many advertisements and other representations of mobile Internet. Interestingly, the image in Figure 7.3 is strongly grounded in mobile phone culture. While mobiles were most often thought—in Western societies especially—to be individual, personal devices, they were shared from their inception (Crawford and Goggin 2008; Weilenmann and Larsson 2001). With the advent of camera phones, especially preceding the easy sending or broadcasting of pictures taken, mobiles were often used to display and show photos to friends, intimates, colleagues or strangers. The phone itself functioned as a photo album, but also a prized repository for digital memorabilia and collectables. In the Chinese context, Leopoldina Fortunati and Shanhu Yang have described such a socio-technical ensemble, not so much as "networked individualism" (Miyata et al. 2005) but as "semi-socialized individualism" (Fortunati and Yang 2012). With the advent of Facebook, apps and other personal media technologies underpinned by much faster mobile Internet, interoperable applications (for example, photos taken on a smartphone can be easily tagged and uploaded to the Internet, with sites such as Facebook or Flickr integrated into Apple iPhone or other applications), and the interweaving of mobiles and Internet in many other ways, sharing increasingly occurs across devices and platforms, reframing the previous centrality of mobile devices to collective, reciprocal investments—and generalizing the cultures of use, and rhetorics, based on sharing.

Sharing of files and content was the defining feature of peer-to-peer (p2p) networks from the late 1990s that came to general notice with Napster (music sharing) and bittorent (Internet downloading of television and movies). Since the appearance at least of Web 2.0 as a notion (O'Reilly 2005), there has been much discussion of the important role of the user in digital technologies, and the requirement for participation in their design and operation. The idea of "user-generated content" accentuated the role that users played in creating the kind of content that became widespread and attractive in digital participatory culture—for example, creating videos for YouTube (Burgess and Green 2009), or play an important role in cocreating online games. Discussions of social media also note the importance of users contributing information and content, and engaging in interaction, as what animates applications such as social networking systems, blogs, or microblogging software such as Twitter. In a corrective to the postulate of the user as ceaselessly productive, some scholars have also noted the importance of the bulk of other users who make sense of such user-productions by their lurking, listening and consumption. However, there is a deeper, much more fundamental logic at stake here. This has become evident in the concerns felt by many about the compulsory nature of many social media platforms that compel users to provide their information.

Facebook is perhaps the most prominent example (not least because it is a repeat offender), raising many issues of privacy and control of users' information. More recently, the rising popularity of locative mobile media has heightened such anxieties. Smartphones, especially, have the capacity to gather detailed information about a user's whereabouts and location, tracking their journeys through space and time (Wilken and Goggin 2014). In addition, apps such as Foursquare are available that present users with the dilemma about where, to whom and why to allow information about their location to be relayed to other users (de Souza e Silva and Frith 2012). Critically, these developments in mobile Internet center on sharing.

As I have mentioned, sharing has been a prominent theme in Internet culture, stretching back past p2p networks to newsgroups, and much earlier sociotechnical developments in the Internet. At a certain point in the development of the Internet, such sharing was theorized in very interesting ways via the notions of the gift (see, for instance, Barbrook 2005; Veale 2003). Yet sharing takes on new dimensions with mobile Internet. The socius of mobile Internet valorizes sharing—as represented in the Movistar advertisement above. Yet forced sharing of information is, at another level, a condition of entry into mobile Internet, and a requirement of participation (Meikle 2014). So there is evidently a tangled mix of practices, ideologies, desires and materialities fused in this emergent social imaginary of mobile Internet, to be confronted and disentangled.

CONCLUSION

It is not possible to offer a full inventory of sharing practices as they figure in mobile Internet. As my brief discussion here entails only some of the most obvious examples that have figured prominently in North America and Europe—especially the former when it comes to locative media. There is some research on sharing practices with mobile phones in a range of other locations and cultures, but as yet little work on mobile Internet. Yet we are aware that there is a very wide diversity of practices developing, specific to particular regions and setting (not least Latin America, with its reliance on mobile broadband).

It is still arguable where sharing is the conceptual wellspring feeding into new social imaginaries in which mobile Internet figure—as this may not be the case across specific locations. That said, sharing certainly features in discourses across different parts of the world, and appears to be at the heart, mutatis mutandis, of new media forms. Of course, sharing has long been believed to be at the heart of human society, bound up with reciprocity, the gift, and communication (Mauss 1966; Strathern 1998). So in mobile Internet we may see an ancient fundamental being invoked in new ways, offering a new solution to the modern problem of how to reconcile individualism, the corrosiveness of the new economy, and what creates the public and private spheres, and community.

This is worth bearing in mind as we survey the global scene in which mobile Internet is profoundly implicated in the epochal social transformations of our time—the rise of China, and its urbanization; the "Latin American" century; the realignments in Europe; the new prospects of Africa; democratic uprisings in the Middle East; the global finance crisis and the dissent it has provoken. Here we have a protean technology, playing a concrete, yet highly resonant symbolic role in the remakings of these societies. We might recall Taylor's remark that "modernity is also the rise of new principles of sociality," and in the centrality of sharing in mobile Internet, and broader convergent digital media, we may yet find something that genuinely productive, that exceeds the harvesting of consumer and users in these new systems of extracting and exploiting value.

SOURCES

Andrejevic, M. *iSpy: Surveillance and Power in the Interactive Era*. Lawrence: University Press of Kansas, 2007.

Andrejevic, M. *Infoglut: How Too Much Information Is Changing the Way We Think and Know*. New York: Routledge, 2013.

Ball, J. "Angry Birds and 'Leaky' Phone Apps Targeted by NSA and GCHQ for User Data." *Guardian*. January 28, 2014. Accessed February 14, 2014. www.theguardian.com/world/2014/jan/27/nsa-gchq-smartphone-app-angry-birds-personal-data

Balsamo, A. *Designing Culture: The Technological Imagination at Work*. Durham: Duke University Press, 2011.

Barbrook, R. "The Hi-Tech Gift Economy." *First Monday*. no. 3 (2005). Accessed March 30, 2014. http://dx.doi.org/10.5210/fm.v3i12.631 (update of article first published in 1998)

Barnes, S., and S. Huff. "Rising Sun: iMode and the Wireless Internet." *Communications of the ACM*. no. 46 (2003): 79–84.

Beaumont, P., and H. Sherwood. "Egypt Protesters Defy Tanks and Teargas to Make the Streets Their Own." *Guardian*. 28 November, 2011. Accessed March 21, 2014. www.guardian.co.uk/world/2011/jan/28/egypt-protests-latest-cairo-curfew?INTCMP = SRCH

Bennett, T., and C. Healy (eds.). *Assembling Culture*. London: Routledge, 2011.

Bingham, H. *Lost City of the Incas: The Story of Machu Picchu and its Builders*. New York: Duell, Sloan and Pearce, 1948.

Boddy, W. *New Media and Popular Imagination: Launching Radio, Television, and Digital Media in the United States*. New York: Oxford University Press, 2004.

Boyd, D., and N. Ellison. "Social Network Sites: Definition, History, and Scholarship." *Journal of Computer-Mediated Communication*. no. 13 (2007). Accessed March 21, 2014. http://jcmc.indiana.edu/vol13/issue1/boyd.ellison.html

Burgess, J., and J. Green. *YouTube: Online Video and Participatory Culture*. Cambridge, UK, and Malden, MA: Polity, 2009.

Crawford, K., and G. Goggin. "Handsome Devils: Mobile Imaginings of Youth Culture." *Global Media Journal*. no. 1 (2008). Accessed March 21, 2014. www.hca.uws.edu.au/gmjau/archive/iss1_2008/crawford_goggin.html#_edn1

de Souza e Silva, A., and J. Frith. *Mobile Interfaces in Public Spaces: Locational Privacy, Control, and Urban Sociability*. New York: Routledge, 2012.

Douglas, S.J. *Listening In: Radio and American Imagination*. Minneapolis: University of Minnesota Press, 2004.

Farman, J. *Mobile Interface Theory: Embodied Space and Locative Media*. New York: Routledge, 2012.

Flichy, P. *The Internet Imaginaire.* Cambridge, MA: MIT Press, 2007.

Flores-Roux, E., and J. Mariscal. "Oportunidades y Desafíos de la Ancha Móvil [Opportunities and Challenges of Mobile Broadband]," *Actos de la V Conferencia ACORN-REDECOM*, Lima, 19–20 May, 2011. Retrieved November 24, 2011. www.acorn-redecom.org/papers/2011Flores-Roux_Espanol.pdf

Fortunati, L., and S. Yang. "The Identity and Sociability of the Mobile Phone in China." In *Mobile Communication and Greater China*, edited by R. Wai-chi Chu, L. Fortunati, P.-L. Law and S. Yang, 143–157. New York: Routledge, 2012.

Funk, J. L. *The Mobile Internet: How Japan Dialed Up and the West Disconnected.* Pembroke, Bermuda: ISI Publications, 2001.

Goggin, G. *Cell Phone Culture: Mobile Technology in Everyday Life.* London and New York: Routledge, 2006

Goggin, G. "Assembling Media Culture: The Case of Mobiles." *Journal of Cultural Economy*. no. 2 (2009): 151–167.

Goggin, G. *Global Mobile Media.* London and New York: Routledge, 2011a.

Goggin, G. "Ubiquitous Apps: Politics of Openness in Global Mobile Cultures." *Digital Creativity*. no. 22 (2011b): 147–157.

Gordon, E., and A. de Silva e Souza. *Net Locality: Why Location Matters in a Networked World.* New York: Wiley, 2011.

Helyar, V. "Usability of Portable Devices: The Case of WAP." In *Wireless World: Social and Interactional Aspects of the Mobile Age*, edited by B. Brown and N. Green, 195–206. New York: Springer-Verlag, 2001.

Hjorth, L. *Mobile Media in the Asia Pacific: Gender and the Art of Being Mobile.* London and New York: Routledge, 2009.

Humphreys, L. "Mobile Devices and Social Networking." In *After the Mobile Phone? Social Changes and the Development of Mobile Communication*, edited by M. Hartmann, P. Rössler and J. Höflich, 115–130. Berlin: Frank & Timme, 2008.

Huston, G. "TCP in a Wireless World." *Internet Computing, IEEE*. no. 5 (2001): 82–84.

Ito, M. "Introduction: Personal, Portable, Pedestrian." In *Personable, Portable, Pedestrian: Mobile Phones in Japanese Life*, edited by M. Ito, D. Okabe and M. Matsuda, 1–17. Cambridge, MA: MIT Press, 2005.

Kirkpatrick, G. *Computer Games and the Social Imaginary.* Cambridge, UK: Polity, 2013.

Kopomaa, T. *The City in Your Pocket: Birth of the Mobile Information Society.* Helsinki: Gaudeamus, 2000.

Kumar, V., S. Parimi and D. P. Agrawal. "WAP: Present and Future." *Pervasive Computing*. no. 1 (2003): 79–83.

Le Doeuff, M. *L'Étude et le rouet.* Paris: Editions du Seuil, 1989.

Maitland, C. F., J. M. Bauer and R. Westerveld. "The European Market for Mobile Data: Evolving Value Chains and Industry Structures." *Telecommunications Policy*. no. 26 (2002): 485–504.

Mariscal, J., Gamboa, L. and Rentería Marín. "The Democratization of Internet Access through Mobile Adoption in Latin America." In *Routledge Companion to Mobile Media*, edited by G. Goggin and L. Hjorth, 105–113. New York: Routledge, 2014.

Martínez, G. *Latin American Telecommunications: Telefónica's Conquest.* Lanham: Lexington Books, 2008.

Mauss, M. *The Gift: Forms and Functions of Exchange in Archaic Societies*, reprint ed, translated by I. Cunnison. London: Cohen & West, 1966.

Meikle, G. *Social Media and Sharing.* Professorial Inaugural Lecture. February, 26, 2014. CAMRI, University of Westminister.

Middleton, C., and J. Given. "The Next Broadband Challenge: Wireless." *Journal of Information Policy*. no. 1 (2011): 36–56.

Miyata, K., B. Wellman and J. Boase. "The Wired—and Wireless—Japanese: Web-phones, PCs and Social Networks." In *Mobile Communications: Re-Negotiation of the Social Sphere*, edited by R. Ling and P. Pedersen, 427–449. London: Springer, 2005.

Morley, D., and K. Robins. *Spaces of Identity: Global Media, Electronic Land-scapes, and Cultural Boundaries*. London and New York: Routledge, 1995.

Natsuno, T. *The i-mode Wireless Ecosystem*. New York: John Wiley & Sons, 2003.

Nokia. *WAP White Paper*. Helsinki: Nokia, 2000.

O'Reilly, T. *What Is Web 2.0: Design Patterns and Business Models for the Next Generation of Software*. 2005. Accessed November 24, 2011. www.oreilly.com/pub/a/oreilly/tim/news/2005/09/30/what-is-web-20.html

Qiu, J. L. *Working-Class Network Society: Communication Technology and the Information Have-Less in Urban China*. Cambridge, MA: MIT Press, 2009.

Quadir, I. "The Power of the Mobile Phone to End Poverty." *TED Talk*. July 2005. Accessed March 6, 2014. www.ted.com/talks/iqbal_quadir_says_mobiles_fight_poverty

Ramos, S., C. Feijoo, J. Perez, L. Castejon and I. Segura. "Mobile Internet Evolution Models: Implications on European Mobile Operators." *Journal of the Communications Network*. no. 1 (2002): 171–176.

Shenker, J. "Egyptian Protesters Are Not Just Facebook Revolutionaries." *Guardian*. January 28, 2011. Accessed November 24, 2011. www.guardian.co.uk/world/2011/jan/28/egyptian-protesters-facebook-revolutionaries

Spurgeon, C., and G. Goggin. "Mobiles into Media: Premium Rate SMS and the Adaptation of Television to Interactive Communication Cultures." *Continuum*. no. 21 (2007): 317–329.

Strathern, M. *The Gender of the Gift: Problems with Women and Problems with Society in Melanesia*. Berkeley: University of California Press, 1998.

Taylor, C. *Modern Social Imaginaries*. Durham: Duke University Press, 2004.

Toyama, K. "Can Technology End Poverty?" *Boston Review*. Nov/Dec 2010. Accessed March 6, 2014. http://new.bostonreview.net/BR35.6/toyama.php

Veale, K. "Internet Gift Economies: Voluntary Payment Schemes as Tangible Reciproc-ity." *First Monday*. no. 12 (2003). Accessed March 30, 2014. http://firstmonday.org/ojs/index.php/fm/article/view/1101

WAP Forum. *WAP White Paper: Wireless Internet Today*. Mountain View, CA: WAP Forum, 2000.

Webster, F. "The Information Society Revisited." In *Handbook of New Media: Social Shaping and Social Consequences of ICTs*, edited by L. Lievrouw and S. Livingstone, 22–33. Thousand Oaks: Sage, 2002.

Weilenmann, A., and C. Larsson. "Local Use and Sharing of Mobile Phones." In *Wireless World: Social and Interactional Aspects of the Mobile Age*, edited by B. Brown and N. Green, 92–107. New York: Springer-Verlag, 2001.

Wheeler, D. *The Internet in the Middle East: Global Expectations and Local Imagi-nations in Kuwait*. Albany, NY: State University of New York Press, 2006.

Wilken, R., and G. Goggin (eds.). *Locative Media*. New York: Routledge, 2014.

Wilson, R. and W. Dissanayake (eds). *Global/Local: Cultural Production and the Transnational Imaginary*. Durham, NC: Duke University Press, 1996.

Yu, H. *Media and Cultural Transformation in China*. London and New York: Routledge, 2009.

8 Future Archaeology
Re-animating Innovation in the Mobile Telecoms Industry

Laura Watts

"iWatch to save Apple?" the banner blazed on the news website in February 2013 (Herzog 2013, n.p.). As an ethnographer of the future in high-tech industry, and an ethnographer of mobile telecoms futures in particular, I read on. A wristwatch-like device with mobile phone capabilities was being heralded as the innovative product that would return Apple to its former share-price glory. The article mentioned that this was not the first such smartwatch, "Sony's original watch" was launched in 2010, "but it failed to catch on with customers" (Herzog 2013, n.p.).

I shook my head, speechless, my spine prickling with déjà vu. For my first encounter with a wristphone had been back in 1998, at the massive consumer electronics fair, CeBIT, in Hanover. In a tall, spotlit perspex case I had admired the chunky black, red and green Swatch Talk, a voice-only watch mobile phone launched by Swatch Telecom (this was in the days before mobile data and the wireless Internet). Twelve years earlier than the 'original' Sony watchphone. Twelve years. Over a decade was missing; the wristphone appeared to have been dead and buried and born again as a 'new' technology, all very convenient for the making of a potential innovation and newness in the mobile phone industry. The headline might have been rather different had it noted that the iWatch was a mere iteration, another watchphone in a product category with a history going back over a decade, well before Apple was even in the mobile phone business.

Lest you pass on, uninterested in the market failure of the obscure wristphone or the market success of a well-known corporation, let me provide you with some further curious artifacts from this archaeology of the future. The wrist personal communicator from Philips was industry news in 1999. Samsung showed their watchphone in 2000. Motorola launched their version in 2001. Just to make the point, in 2004 Seiko announced the world's first wireless watch, at exactly the same industry conference as a Thinking Materials phone-watch and a gaming wristphone from Teleca, and, if you are still unmoved, LG was touting their Touch Watch Phone in 2009, only one year before Sony's 'original' wristphone mentioned in the news report.

It would appear there has been ongoing development in the product category of wristphones over the past fifteen years. From where comes the newness and innovation that is proclaimed to be the salvation of a high-tech company? To provide some perspective, the Wi-Fi standard, now so common that the air in many parts of the world is always-on with wireless Internet, that taken-for-granted standard is more or less the same age as the watchphone; they have both seen the same number of years of development (the 802.11 standard was first published in 1997).

In the normative version of technological innovation, the version that underlies the news report, technological development is supposed to be an increasing, linear or step-wise progression; things can only get better, so to speak; the 'new' is always more than the 'old.' And shareholders will hold you to that version, as the report makes clear; this is not an idle theoretical point. The iWatch was new; it was proclaimed to be an exciting future for the mobile telecoms industry. But that future could only be 'new' if all those former 'new' and world-first watchphones were made absent, forgotten and buried. Any connection between those old futures and the new future in 2013 had to be severed; no relation made. All the old futures had to be quietly killed off.

I felt like the protagonist in Bruno Latour's book on the Aramis transport system, another technological future that had died a death (Latour 1996). Latour's detective character was on the case of the murder, asking: who killed Aramis? I had a feeling that some of the answers to that question, which drew upon a sociotechnical and actor-network theory approach to techno-logical research and development, would be relevant to my own murder investigation. I wanted to ask: was this a one-off murder, or was the killing of the future widespread? How did the murders take place, how did those futures die? And were their deaths really necessary to making innovation in the mobile telecoms industry?

This was why my spine prickled with déjà vu and the desire for action. As an ethnographer, I had evidence of those former watchphones erased from existence in the news report, and I felt compelled to do what every ethnog-rapher must: make a fieldsite.

FIELDSITE FOR UN-DEAD FUTURES

A fieldsite is not a part of the world out there waiting to be bagged and tagged in an ethnographic notebook. As has been well argued, a fieldsite is an effect of deskwork and fieldwork; it has to be made, woven back and forth between the located analyses of an ethnographer at their desk, and their particular, partial experiences of the people and places they collaborate with in the field (Strathern 1991; Gupta and Ferguson 1997). As an ethnog-rapher of technoscience, with a former career as a designer in the mobile telecoms industry, my knowledge and collaborations are located in ways that need a slower moment to explain.

There is a particular scent to futures. I tend to smell them, catch their scent on the wireless ether. It is a sensitivity I have as an ethnographer, which has been developed over several years. The mobile phone network standard '3G' was a future once, a prophecy made in industry predictions and then in corporation-breaking billions, and it was where I began my fieldwork inside the mobile telecoms industry in 2003 (Watts 2008). For four years I variously inhabited the design studio of a major mobile phone manufacturer and other industry sites in the UK. But those experiences are also entangled with five prior years enmeshed in mobile phone design and development during a former career, 1997–2001. Through these experiences my senses have become attuned to future-making in the industry. My fieldsite is a knot that ties these experiences together.

Given that this fieldsite (and any fieldsite) is only ever a fragment, as I write I elide 'the' mobile telecoms industry with these fragments. But, of course, a mobile phone operator makes a different future than a manufacturer, and then there are differences over geographies, the mobile phone world in North America is rather different to Europe and Asia, and so on.[1] Just to make my generalization explicit.

I have a nose for futures, I said, and by 'futures' I do not mean some unknown future over the temporal horizon, nor am I interested in reading the statistical entrails or the augury of future studies or futurology (although that is a serious business). The futures whose scent, fresh or decaying, I can but follow, are those made around me; this is why I speak of 'future-making'. Futures are made and fixed in mundane social and material practice: in timetables, in corporate roadmaps, in designers' drawings, in standards, in advertising, in conversations, in hope and despair, in imaginaries made flesh (Brown et al. 2000; Bloomfield and Vurdubakis 2002; Jensen 2005; Rosenberg and Harding 2005; Adam and Groves 2007). Following Haraway's situated knowledges, I think of these as situated futures; for futures are both material and semiotic (Haraway 1991). They are made in practices, with things such as standards and strategy documents.

So, since they are situated and made by particular people in places, with all the social and technical relationality that implies, their making can be recorded, evidence gathered and my fieldsite constituted.

Consider this fieldsite as the making of an archaeological excavation, unearthing the remains of dead (and perhaps not so dead) futures in the mobile telecoms industry. So let me dig, let me excavate my fieldsite and expand the trench that began with the wristphone.

There is already something in the ground, round and smooth . . .

PEBBLE FUTURES

In another article on the iWatch, there is a comparison with the 'Pebble watch.' This accessory can display notifications from your mobile phone on its curved screen wrapped around your wrist; one reviewer called it "the first smartwatch for regular people" (Patel 2013). Not a world-first, then, but still a first. Yet reading this review, it's not the reanimation of the wristphone future that blooms hot and fetid in my nose, it's that other word:

Pebble.

Back in 1997 I had helped to design a pebble-shaped mobile phone concept. The shape had been so iconic that I had even written a short science fiction story for a company newspaper featuring a pebble wireless device. But when I returned to the industry as an ethnographer, seven years later, the salty stench of pebbles was still present.

During my fieldwork, I remember the company minibus driving me into the clean, chlorinated white and green buildings of the research and development campus wherein lay a mobile phone manufacturer's design studio. This was in the Thames Valley high-tech zone to the west of London, a walled mass of IT and telecoms corporations, pressed up against Heathrow airport. My hard-won visitor's pass let me pass into only one of the three buildings, then into an elevator and up four floors. It green-lit my access through a frosted glass door and into the off-limits design studio, a sanctum few employees were permitted to enter. This hidden world had a familiar aesthetic: Herman Miller chairs, birch desks, science fiction prints on the walls (a character from the film *Blade Runner* was one I immediately recognized), as well as project rooms, their corkboard walls covered, floor to ceiling, with notes, drawings, illustrations and torn-out magazine pages.

In the central 'hub' room, a team of industrial designers sat in *ad hoc* Ikea chairs, hunched over pens and pads of paper. They were discussing the design trend for a handset to be launched in two years time. A future was being made, so I took out my notebook (no other recording devices were permitted on such hallowed ground). Here's an extract of the conversation, as I noted it:

"Ecological, in a material sense, [means] is natural."

"Choice of natural has integrity, do it where need for flexibility has a rationale."

"Stone is more natural than white . . . White is ageless . . ."

"Products [need to] look like they are in motion . . . Pebble shape has motion."

There: pebble.

The meeting had ended and one of the designers invited me to his desk and handed me a polished, rose quartzite pebble. He explained that it had inspired his design for a mobile phone to be launched next year. The pebble was a tactile mnemonic for the sensations he wanted to evoke (Watts 2005).

Caught in my notebook and then in my hand: the fixation on a particular shape, a particular sensation, a particular future.

And the fixation remained during my fieldwork. Two years later, in 2006, Motorola launched the PEBL mobile phone. Its accompanying advertising campaign, "Shaped by Nature," featured the origin myth of the PEBL: a meteorite flung from space in some distant past, to fall onto a beach, to be pounded and smoothed over aeons by the sea, wind and rain, to be at last picked out of the waves by a barefoot user—a creation myth more akin to King Arthur's sword, Excalibur, than to technological innovation or invention.

And so we return to 2013 and the Pebble watch. My evidence traversed sixteen years in total, sixteen years in which the Pebble remained always future, never past. In the trench, then, buried in the dirt of my fieldsite were not just the corporeal remains of long dead wristphone futures, but also long dead pebble phone futures.

The erasure of history and the lack of corporate memory in high-tech industries is an old theme, however, and has been remarked on before. For example, Brian Schiffer's excavation of the portable music player and its many forgotten histories suggests that the reworking of high-tech history is a strategic, political act by corporations intent on ownership of both past and future (Schiffer 1991). Kim Sawchuk makes clear the white middle-class politics embroiled in the development of wireless technology and its happy-family histories (Sawchuck 2010). Whose future gets made, and whose gets erased and buried, is an important question. Futures are situated, as I said; they are always located in socio-cultural epistemology, they always have race and gender specificities.

Given that burying and forgetting technological pasts is not surprising, then, given that quietly 'bumping off' the previous technological generation is not uncommon, why remark upon it? What does digging around in this fieldsite do that is interesting? Merely naming the dead futures in mobile telecoms, such as wristphones and pebbles (and there are more), is not enough.

There is one point that needs to be made.

The same industrial designers who were reanimating the sixteen-year-old pebble phone future also claimed: "It feels as if we're always under-predicting . . . on the backend of every curve . . . It's shocking the speed at the moment." Let us be clear, this is an industry whose tropes, whose stories, are about speed and high-speed technological change. Speed is a story with extraordinary effects: the news report on the iWatch is suffused with the perceived failure of a company to reproduce this story, and the direct correlation with its dwindling share-price. "Innovate or die" is the oft-repeated business mantra.

Making this fieldsite of long-dead and reanimated futures is a clear counter to the mobile telecoms industry hyperbole of high-speed development.

Digging up and attending to reanimated, zombie-like futures claimed to be 'new' calls into question what counts as innovation, as named by the industry. It makes clear that the work being done in making pebble and wristphone futures is not limited to the invention of a 'new' product or prototype; there is much more going on. Devices such as the iWatch and PEBL enact and reproduce already existing futures. Within this industry, and perhaps other similar industries, there are not changing futures, but enduring futures. Industry innovation maintains the same future over considerable periods of time. Its practices, prototypes and products hold a future steady, whilst it simultaneously maintains a trope of 'shocking' speed (and there is perhaps much to say about that as a low-risk strategy in a high-risk game).

Social studies of science and technology have long critiqued normative versions of high-speed technology invention and innovation. Now-classic texts have discussed the *deus ex machina* problem with innovation as technological determinism, and demonstrated how technology does not appear fully-formed or derived from itself, but is an effect of extraordinary and difficult sociomaterial labor and relation-making (Latour 1987; Bijker et al. 1989; Bijker and Law 1992). More recently, much work, also in anthropology, has nuanced the difference between innovation, invention and creativity and how these are done in different places and have diverse effects, such as in the 'technological society,' which is defined by its political reliance on technical change (Barry 1999, 2001; Ingold and Hallam 2007; Nowotny 2008). Lucy Suchman, in particular, has explored how high-tech industry innovation is often, when looked at in everyday practice, much more a matter of artful integration and local improvisation, involving the reconfiguration of existing relations between people and things (Suchman 2002, 2007), and these are just some pertinent examples from a very extensive oeuvre. So innovation is much more interesting than the sudden appearance of a world-first wristphone (again). But poking a stick at all the dead futures that have been buried along the way in the mobile telecoms industry does not get my fieldwork much further than these existing critiques.

Inside the industry, the notion that a future, that the shape of things to come, might be decades old, that technological innovation might take considerable time, decades even, and be fraught with long-term failures and artful resurrections, is not tenable. It cannot be made visible in an industry whose salaries and company existences, both large corporations with shareholders and small start-ups with venture capital funders, depend on a version of innovation that is about not just newness, but newness-at-speed. I might go as far as to say there is incommensurability between this industry innovation and the version created by the critical academic work I have cited. The mobile telecoms industry enacts innovation through the efficient murder of its former darlings, so that the 'old' can be made 'new' again, which is a very efficient way to satisfy the requirement to 'innovate or die.'

In comparison, sociotechnical critiques of innovation are concerned with how the 'new' is made, by making the 'old' visible. These do not go together. They are different kinds of work. So, aside from repeating the critical move to make the 'old' visible, what else is there for an ethnographer of technoscience to do?

I need to keep digging, keep following my nose for rotting futures. But I also need to attend to the partiality of these remnants, for dead futures are not whole bodies but partial assemblages (as any technology must be). The pebble phone and wristphone are Frankenstein monsters, made of bits and pieces, bits of things and people put together; they are sociotechnical relations that have been plugged in and switched on (Law 1991). What I can do is ask more questions: what relations constitute these futures? How are these futures made, and how did they die?

Back to the trench, then, and there is already a long, thin thread I can tug from the dirt . . .

UBIQUITOUS FUTURES

The thread I began to pull from the dirt was an old, familiar line to me. An industry magazine described it as "a runaway train, roaring down a path to disaster, picking up speed at every turn, and we are now going faster than human beings can endure" (Malone 2003, front cover). It was perhaps one of the most famous versions of newness-at-speed in the high-tech industry: the line known as Moore's Law. This was a prediction made in 1965 by a founder of Intel, Gordon Moore, of the exponential doubling of components on an integrated circuit year on year (Moore 1965). Since then, Moore's Law has been taken to stand for the prediction of an exponential annual increase in almost any measure in the industry, from bandwidth to battery life. The manager of the design studio in the Thames Valley called it "the hockey stick effect" (reflecting the shape of the line on a graph), and was as frustrated by its stranglehold on company decision-making as the magazine journalist.

This speeding thread had endured in the industry for at least half a century, far longer than wristphones and pebble phones. Its age reeked in my sensitive nostrils, an ancient sinew. I bent closer, pulled on the thread, seeking ethnographic evidence for how it had endured, and found it was entangled in endless exponential graphs.

The graphs took me from the Thames Valley to Cannes in the south of France, famous for its film festival, famous in the mobile telecoms industry as the former location for the massive event known as the 3GSM World Congress. I had attended this industry conference along with thirty thousand other delegates in 1994, an event so vast that it had increased the local population by 50 percent. (By 2013, it had relocated to Barcelona and there were seventy-two thousand delegates swarming the city).

I remembered that the company bus had rolled through the palm-lined strip of La Croisette, its famous beach promenade. I had watched women in furs trailing small dogs, crisp black-and-white-uniformed waiters serving coffee to dark-suited business men (and they were all men), the Versace and Armani boutique shops, and the gleaming white hulls of corporate-sponsored yachts packed into the marina, or what the industry journalist I interviewed called "the floating gin palaces." And hanging from every luxury hotel façade, every street pole, pasted to almost every vehicle, was industry advertising. Even the sea and air were branded through evening fireworks that blazed with logos, reflected in the water. That morning the air was magnificently bright to my British eyes, as the venture capitalist said, "the feeling I always get in Cannes when the sun's coming up and it's in February and it's the first time any of us have seen sun for six months in Britain is: sunshine, optimism."

The venture capitalist and journalist had passed on their experience as well as the conference pack as part of my fieldwork in 2004. The conference pack contained the PowerPoint slides from the presentations, and they were full of hockey-stick-shaped graphs. The conference speakers had made countless exponential, visual predictions: music revenues would increase exponentially from 2003 to 2008, subscribers for location-based applications would increase exponentially (no need for an axis measure on that one), the number of 3G mobile phone models over a year would also see exponential growth, revenue from mobile media messaging (sending photos between phones) was another hockey stick drawn from 2003 to 2008, its rise only surpassed by revenues for mobile instant messaging (not to be confused with text messaging) during the same period.

I could go on.

No matter the y-axis or the time along the x-axis, the shape was resolute, repeated until it was mantra: the speed of the industry was increasing, in every measure. The future was acceleration. This was how the runaway train roared down its path to apparent disaster. This was how newness-at-speed was done, in large part. Here, the line of Moore's Law was reanimated into near-mythic proportions; how could such an enduring future, re-enacted over and over by conference speakers from around the industry, be questioned?

Now I knew something of how this monster was assembled.[2] But how else was this zombie future made so potent?

There were two additional, already well-documented, practices.

First, as I have explored, technological determinism was endemic in the industry. One mobile telecoms company CEO at the conference said in a news report that: "[mobile telecoms network] operators are frightened to death of technology. It comes at us like missiles" (Ee Sze 2005). That fear of technology was an effect of the normative separation of the social and technical. The industry erased itself, its social organisation and people, from its technology innovation, and so erased its potential censure (and responsibility). Without such separation it's hard for a mobile phone designer or CEO to argue that the speed of change is shocking, for they are the ones making the change.

Second, was that well-studied technoscience power, the power of number, and trust in numbers (Porter 1995). The exponential lines on the PowerPoint slides adopted the apparatus of fact-making, the graph, to evoke trust in their numbers, and trust in the prediction of the line. Failure to label the axes was not an issue since there was no science being done, only a simulacrum that evoked measurement and scientific practice. The graph was read as reliable and trustworthy as a future, because it was a graph. The slides did not reanimate the name of Moore's Law; it was not the label that mattered. They reanimated the graph from Gordon Moore's article, the hockey stick curve. They reanimated a mathematical shape, reproducing a prediction that had apparent mathematical certainty, and there lay a considerable part of its power.

But this monstrous line had a third potency: it was an asymptotic curve, as mathematicians might note. It was tending ever-upward toward a future, infinitely far on the horizon. This accelerating line had some impossible-to-attain end point, a future vision that the line and the industry yearned for. The advertising from the powerful industry association, who organised the conference in Cannes, stated that vision in an advert in the conference daily newspaper: "GSM grew from a vision. A revolutionary vision that mobile phones should keep customers connected anytime, anywhere, even when crossing borders" (*GSM Daily*, February 25, 2004). The word that encapsulated this vision, which was on so many PowerPoint charts at the conference, on my tongue in 1994 and still on the industry tongue ten years later, was "ubiquity."

I asked the industry journalist for a definition. He put it simply: "[It's] everywhere you go in the universe, and everywhere you go it works the same way."

Everywhere.

Anyone Anytime Anywhere, to complete the well-used triptych in the industry. This was the industry vision, the tendency of the exponential line, and the dream that was evoked, sometimes explicitly, by all those graphs. Ubiquity was made powerful through its universal, all-encompassing vision. Nothing seems outside of ubiquity. It seems weightless, hovering nowhere and everywhere.

But I said earlier that futures are always situated in epistemology, they are always located and have politics. So the knee-jerk question is always to ask: Whose dream is this? It seems to be a dream for everyone and anyone; who would not want to be connected anytime, anywhere? But it is a classic god-trick, that is, knowledge and knowing that appears to be omnipotent, for only a deity could be everywhere in the universe (Haraway 1991). In practice, telecommunication networks are necessarily much more specific (and more interesting). They are sociotechnical infrastructures that are always partial and patchy in their implementation (Star 1999), with not-spots where there is no mobile phone signal, and with people who are not important enough, or not wealthy enough, to warrant the cost of the infrastructure; 'not-ones' who are not included in the *anyone* named by the dream, such as those living in places remote from cities. So this future is neither weightless nor universal. It is sociomaterial, with absences and presences. Ubiquity is not some special class of über-future, an unquestioned good over and for all; ubiquity is particular.[3] Ubiquity is just a particular mobile telecoms future, constituted by people in places, as much as the wristphone and the pebble phone.

So ubiquity was taken-for-granted as a future in mobile telecoms, its unquestioned status derived from technological determinism, from trust in numbers, and from a god-trick.

LANDSCAPES OF UN-DEAD FUTURES

Through my fieldsite I had excavated three futures, now, that the mobile telecoms industry had reanimated and made to endure: wristphones, pebble phones and ubiquity. But I was not content to just count and name the bodies, as I said. As an ethnographer of technoscience I was looking for the parts that constituted these monsters, and that must include the soil; in archaeological terms, the stratigraphy, or the location.

The dirt in which all these futures were buried smelled the same to my future-sensitive nose, and that was very interesting.

I have a refined nose for futures, which perhaps does not parse in text. So let me explain the quality of their smell.

The dank smell of situated futures rose up from the dirt, literally from the place where they were made. This is because my professional senses have been honed via an attention to the landscapes where knowledge gets made. Tim Ingold was a key instructor in this attention: his classic work on the temporality of landscape argues that as we walk and move through the world we perform its memory, "we know as we go, not before we go" (Ingold 2000, 230). David Turnbull builds on this to argue that particular movements through particular landscapes perform particular knowledges (Turnbull 2002). Just to be clear, this is not environmental determinism, place does not determine knowledge, rather I regard place as an actor in knowledge-making; the heterogeneity of actor-networks is merely expanded beyond the social and technical (Law 1992). So, since the futures of which I speak are also made by things and people moving through places, my senses have been tuned to the dirt under their feet. Particular dirt gives rise to particular futures; that is, where the designers, pebbles or CEOs go affects what they know and the futures they may weave.

So when I speak of the 'smell' of situated futures, I mean that it has particular qualities. There are three qualities to situated futures: the epistemological landscape (the situated knowledge), the socio-technical landscape (the embedded infrastructure) and the geographic landscape (the place). These three are entangled and inseparable in practice, of course.

And, as I said, the smell for all three zombie futures was very similar; they shared similar qualities.

The wristphone was being reanimated, allegedly, from deep within the Apple campus in Cupertino, California. Even though I had never visited, I smelled its normative innovation and middle-class wealth in the industry news reports (Wikipedia cites it as the eleventh richest city in the United States), and in other ethnographic fieldwork on Silicon Valley innovation (Stone 1996; Finn 2001; Suchman 2005). It is a place where the wireless Internet is ever-present, and the mobile phone signal is strong.

The pebble reanimation takes place in another gleaming, secretive, corporate research and development compound, in another high-tech valley, this one in Britain. As with Silicon Valley many of the world's large telecoms and IT companies were here (Microsoft, Oracle, LG, HTC, RIM, Cisco, Cable & Wireless, the list was extensive). This was the landscape where fourth-generation mobile Internet (4G LTE, for the technically inclined) was trialed, well before the network was rolled out to other parts of the UK a year or so later. The air and the ground hum with high-speed broadband. The Thames Valley is topographically flat, and the population is dense with disposable income, ideal for cost-effective and technically-effective mobile infrastructure. Here, you can live as though 'always on.'

Cannes, on the French Riviera, backed by the mountains of the Alpes-Maritimes, may seem to be somewhat different, its situated futures an entirely different scent. But you need to be sure to smell the dirt during the conference, when ubiquity was reenacted once more. The local knowledge was not that of the long-term residents, but rather the knowledge of the thirty thousand industry delegates who flew in for the three days and doubled the population, the majority senior management and board level decision-makers. This was not a place for a marketing junket, but where deals between companies were cut. As a venture capitalist I interviewed said, "from the beginning of the day to the end of the day I have a schedule . . . We did five deals on the first day. You go and do deals." This is normative innovation doing business. The infrastructure was also entirely different. Thirty thousand additional people puts something of a strain on a mobile phone network, so that year (and every year) a team of engineers worked for five months to install an entirely new 3G network in Cannes, involving new antenna on and in buildings, hotel rooms turned into equipment rooms, and so on; a massive social and technical undertaking, and it was temporary, operational only for the conference. All done so that when the mobile telecoms industry turned up, they could just switch on their phones and have mobile Internet as though it were always on, anytime, anywhere, and it seems pertinent to note that I was forbidden access to the conference unless I paid the entrance fee of 3,000 euros (about US$3,800); the cost of its visitors pass restricted access as surely as at the other locations.

All three landscapes had the same smell, epistemologically, infrastructurally and geographically. Three zombie futures (the wristphone, pebble phone and ubiquity) were all remade in places that were remarkably similar in their situated-ness. The industry might be global, but it was not universal and anywhere; it was much more parochial in where it made its home and where it made its futures. The landscapes where the industry lived and dwelled were similar locales. If a place had an unfamiliar landscape then it was remodeled, as the massive re-landscaping of Cannes during the conference demonstrated. It was almost as if the industry lived and thrived in only one particular place, a single archipelago whose islands were connected by airports and air corridors.

It may seem like a rhetorical point: the landscapes of the high-tech industry are islands sprung up like reefs around wartime wireless and computing histories, and around major international airports (there are obvious benefits to this when you are moving large numbers of employees around the world for meetings). But given that landscapes are actors in how futures are made, given that place is not a backdrop, not mere context to the main event in the design studio, it is perhaps important to note that the same landscapes enact the same futures.

And these landscapes are privileged in the future of the mobile telecoms industry; something you know all too well if you happen to be a farmer on a Scottish island, or in other wireless-Internet-disadvantaged sites in the world.

So, landscapes were another part to the zombie futures of the mobile telecoms industry. Futures were made of bits and pieces of place, as much as they were of dead wristphones, design studios, quartzite pebbles and dreams.

No wonder they smelled of dirt and decay.

But were there any other unexpected parts in my dissection of these un-dead futures? Yes, there was one thing more, one additional scent that my nose as an ethnographer could detect. That line, the hockey stick curve of ubiquity, smelled ancient, it smelled old, really old, older even than the industry itself . . .

IMAGINARIES IN THE AGE OF UBIQUITY

How old was ubiquity, then, how far back did the line and its asymptotic ideal go? This might seem the obvious question, but it made the mistake of assuming that ubiquity (or any future of the mobile telecoms industry) was reanimated over and over precisely the same as before. Looking for the original ubiquity was rather like looking for the original watchphone (à la Sony in the news report). You could always point to some earlier object that was similar. This was because the futures that were being resurrected by the industry were not wholes, they were heterogeneous mixtures, monstrous hybrids. Like Bruno Latour's Aramis and Mary Shelley's Frankenstein, they were mash-ups, they were made from bits and pieces: dead bits of ubiquity that had gone before, but also pieces of other things like designers and landscapes. The iWatch enacted the same future as the Swatch Talk, but they were not identical. It was a repeated performance, a rehearsal, always containing the possibility for difference. This was not surprising given that the future was remade in local people and places; if you know as you go, then you know futures as you go, and particularity is everything.

It was a hopeful thought. Things might be otherwise, in time, much more time than the industry presupposed. But time did remain in my nostrils, that decaying smell of age, of the *longue durée*.

I could only dissect some of the parts that comprised the body of this un-dead ubiquity, but even the parts I could name were enough to provoke marvel at the age and endurance of this monster.

There were twenty-first century parts, of course. One made by the industry association who organised the conference: "Imagine a world without wires; a seamless, limitless world of verbal and visual communications," it said. Its newspaper advert spoke of the wireless world, the global mobile Internet, a very contemporary ubiquitous future it seemed.

Still contemporary was the twentieth-century book I was handed during my fieldwork, a book that my industry informant called their "bible on the future." It spoke of a very similar ubiquitous moment in a global communications network, "a wondrous day when electricity would endow the planet itself with cosmic intelligence" (Kaku 1998, 43–44), and this industry bible made the connection between the two explicit: "In the twenty-first century the telecommunications revolution, ignited by the microprocessor and the laser, will finally make Hawthorne's vision come to pass" (ibid.), and now for the putrefaction: this cosmic intelligence was a future envisioned by novelist Nathaniel Hawthorne in 1851. His ubiquity was a nineteenth-century future inspired by the wonders of electricity and the telegraph.

My sense of smell was supported by many others who have dug down into the histories of the wireless world: legendary physicist, Nikola Tesla, envisaged a "World Wireless System" in 1915 (Gabrys 2010); and universal communication is an ideal that has been linked back to Judeo-Christian myths of the Tower of Babel (Mattelart 1999). Ubiquity was a future older than the mobile telecoms industry itself.

Ubiquity was also fiction, and that was interesting to me as an ethnographer. Ubiquity was part science fiction, part biblical myth. The exponential curve on the graph might resemble mathematics, the wireless talk might resemble technoscience, but this zombie future was lumbering around with imaginary parts. That is not to say they were superfluous or ephemeral, quite the opposite. These imaginaries, these fictions were material-semiotic all the way down; see the very material evidence: the book, the words, and the industry association advert.

I should not have been surprised. The mobile telecoms industry was immersed in science fiction; thus the *Blade Runner* character on the wall of the design studio. Another book from my fieldsite, written by a well-known industry futurist, agreed: "If I ever had a dream of mobile communication it was fuelled by my Tuesday night experiences as a student in the 1960s . . . the TV room would be packed with anticipation, people waiting to see James T. Kirk beam down to some unknown planet. His first act was always to confirm safe arrival through his flip-top communicator" (Cochrane 1997, 77–78). When I was working inside the UK industry during the 1990s, the next generation of Star Trek had the effect of turning the company laboratory into a silent wasteland on a late Wednesday afternoon. This entanglement of science fiction and technology innovation has been well-documented, particularly through feminist scholars on cyberpunk and the Internet (Featherstone and Burrows 1995; Balsamo 1996; Bloomfield 2003; Kirby 2010). Innovation has always been made, in part, through imaginaries.

As I followed this scent of science fiction in the futures of the mobile telecoms industry, my trench appeared to be coming full circle. For the news reports on the Swatch Talk, on the LG wristphone and on several others all cited the exact same origin for the wristphone. They all cited Dick Tracy, the science fiction detective comic book character created by Chester Gould in the 1930s, who was famous for speaking into his wrist communicator. Judith Nicolson has examined the racial and political effects of Dick Tracy and his science fictions; there is nothing neutral about this technological imaginary (Nicholson 2008). Science fictions are always located, they are just as situated, just as material-semiotic, as any other future.

There were two things here that I cared about as an ethnographer of technoscience, two things that mattered to me in the imaginaries of the mobile Internet.

First, technology futures in the mobile telecoms industry are made, in part, by science fiction. The zombie futures reanimated by the industry in the name of innovation have social and technical parts, but they also have imaginary parts that are no less influential. There is a tendency to ignore the imaginary in ethnographic studies of technoscience, hive it off to literary or media theorists and retain focus on the oily, mechanical bits and on flesh-warm bodies (although there are always good exceptions, Marcus 1986). But as my case study trench had un-earthed, zombie futures are made of such fictional stuff, with juicy evidence to show. In actor-network-theory ways of making the world, the imaginary was just another actor category, more heterogeneity. It is usual to speak of assemblies of people and things, and the inseparability of the social and technical. It is less usual to speak of the inseparability of people, things, the environment and the imaginary. But in the mobile telecoms industry, futures were made of such stuff.

But the thing that mattered most to me was that science fictions also fix the future into a static shape; they establish conditions of possibility (to draw on Foucault). They were not neutral or innocent but imbued the futures that reanimated them, knowingly or not, with those politics and those conditions. The transport system, Aramis, whose death was studied by Bruno Latour, was fixed and held in stasis for seventeen years by the impossible dreams of those who conceived it (Latour 1996). The conditions for science fiction and dreams are radically different to the hard negotiation, blood, sweat and tears of technology innovation. Fiction has the luxury of impossibility, for science fiction is partial in all the ways that matter to innovation: science fiction does not consider how a technology is manufactured, how is it maintained, how much energy it uses, what is the warranty, what is the intellectual property, who is envisaged as a user, how much it costs, which standards it supports, how is it packaged, installed, recycled, reused; all these things are absent, and all these things need to be agreed upon for a technology to move through a product development cycle. There is no iWatch without the small print and a large infrastructure (or two), all of which Dick Tracy and his author never had to concern themselves with.

So no wristphone can ever be the Dick Tracy dream, no wireless world can ever be as ubiquitous as a global electronic brain or the Tower of Babel, and no mobile phone can ever be as natural as a pebble or as mythic as a meteorite that has fallen to earth. Science fictions are not only unobtainable; their conditions of possibility, if established by the needs of fiction and a good story, may not be relevant futures at all, and yet they remain, binding the future.

Science fiction futures endure in the mobile telecoms industry. As a marketing manager said in the Thames Valley design studio, speaking of a proposed new videophone: "It is the science fiction dream, I cannot believe you never want to do that."

All of which led to an interesting proposition. As an ethnographer of technoscience, who acknowledged that she generated her fieldsite from fragments of evidence, might I write a scientifiction (as Bruno Latour called Aramis as a genre), some empirical science fiction, that might do work in the mobile telecoms industry to un-fix its futures? Perhaps write an intervention into innovation as reanimation?

FINAL THOUGHTS ON A FUTURE ARCHAEOLOGY

Some might argue it is not the role of an ethnographer to write science fiction, or even scientifiction, as an intervention into her fieldsite. Science fiction writers are already writing interventions into the future, and some are attentive to those futures they make present. Writers such Ursula Le Guin, for example, who has explored different futures for sex and gender, and argued extensively for both fantasy and science fiction as an important part of our engagement with the present; what matters is "our thoughts and our dreams, the good ones and the bad ones, and it seems to me that when science fiction is really doing its job that's exactly what it's dealing with. Not 'the future'" (Le Guin 1989, 143). It is a sentiment mirrored by Warren Ellis, who argues that "good science fiction, challenging science fiction, is never about the future we expect. SF has never been about predicting the future. It's been about laying out a roadmap of possibilities, one dark street at a time" (Ellis 2006). It is this attention to the conditions of possibility, to our dreams in the present, that I agree with; for it is in the present that the futures of the mobile telecoms industry are reanimated. As an ethnographer of technoscience I am also attending to these dreams in the present; this entire excavation and fieldsite has been just that. So I share the same care as these science fiction writers, but my method for writing, my toolkit for storytelling, as an ethnographer, is rather different.

So making an intervention is a methodological question, for me. This excavation has done work to provide tools for a method that might intervene in mobile telecoms innovation.

My method must include my commitment to the empirical, to my fieldsite data (however partial). But technoscience knowledge-making has always been generative, knowing the world is always both an empirical and creative process; you always have to fill in the gaps between the evidence, so to speak. The zombie futures I have constituted are made from parts, decaying fragments, which I have written together. Hence why the metaphor and approach of archaeology has been insightful in this chapter. For archaeology is an approach that also acknowledges "the productive and generative potential of breakage and decay," as archaeologist and prehistorian Joshua Pollard argues (Pollard 2004, 60). Both ethnography and archaeology are concerned with generating stories from partial fragments of evidence. But whereas archaeology is concerned with stories of the past, I am concerned with generating stories of the future. Writing stories of the future from fragments of empirical evidence is a method I call Future Archaeology (Watts 2012).

But this chapter has added one further tool to a method for intervention into the industry. The zombie futures of the mobile telecoms industry are constituted by heterogeneous parts, including people, things, places and fictions. In my excavation of three examples of enduring futures in the industry I found that, although people, things and even fictions are various, the places seem to remain remarkably similar across all three. If enduring places lead to enduring futures, then perhaps it is the geography of the mobile telecoms industry, the places wherein it dwells, that are a site for generative interference. If the places where the mobile phone industry lived and moved, remembered and dreamed, were different, then perhaps the futures they designed might be different in some way, too. It is something I have begun to explore elsewhere, but there is much more to be done (Watts 2005, 2008, 2012).

I began this excavation and fieldsite by asking if un-dead futures were really necessary to making innovation in the mobile telecoms industry. Perhaps through a method of Future Archaeology and its attention to place and the role of landscape in high-tech futures, there might be other kinds of innovation that are possible, in other kinds of places. Unfixing the landscape might unfix the future.

I also said I had hope, and so it remains. For innovation as newness-at-speed is already only one version of innovation in the world, as I have explored. Future-making that cares and takes responsibility for the futures it makes is already practiced by many science fiction writers, whose parts constitute high-tech innovation. So within the zombie futures of the mobile phone industry there are already hopeful interventions. Zombie futures are perhaps not the walking dead that I feared, for they are always reanimated with differences, their parts can always change, and others are already tinkering with their innards, and I intended to do a little ethnographic tinkering as well.

NOTES

1. For an example of the mobile telecoms industry in another world region, in this case the Caribbean, see Horst and Miller (2006).
2. For a parallel example see Genevieve Bell and Paul Dourish's work on the future in ubiquitous computing (Dourish and Bell 2007; Bell and Dourish 2011).
3. For a discussion of 'monsters' as an actor-network assemblage see (Law 1991).

SOURCES

Adam, B., and C. Groves. *Future Matters: Action, Knowledge, Ethics*. Leiden: Brill, 2007.
Balsamo, A. *Technologies of the Gendered Body: Reading Cyborg Women*. Durham: Duke University Press, 1996.
Barry, A. "Invention and Inertia." *Cambridge Anthropology*. no. 3 (1999): 62–70.
Barry, A. *Political Machines: Governing a Technological Society*. London: Bloomsbury Academic, 2001.
Bell, G., and P. Dourish. *Divining a Digital Future: Mess and Mythology in Ubiquitous Computing*. Cambridge, MA: MIT Press, 2011.
Bijker, W., T. Hughes, and T. Pinch. *The Social Construction of Technological Systems: New Directions in the Sociology and History of Technology*. London and Cambridge, MA: MIT Press, 1989.
Bijker, W., and J. Law (eds.). *Shaping Technology/Building Society: Studies in Sociotechnical Change*. Cambridge, MA: MIT Press, 1992.
Bloomfield, B. "Narrating the Future of Intelligent Machines: The Role of Science Fiction in Technological Anticipation." In *Narratives We Organize*, edited by B. Czarniawska and P. Gagliardi, 193–212. Amsterdam/Philadelphia: John Benjamin, 2003.

Bloomfield, B., and T. Vurdubakis. "The Vision Thing: Constructing Technology and the Future in Management Advice." In *Critical Consulting: New Perspectives in the Management Advice Industry*, edited by T. Clark and R. Fincham, 115–129. Oxford: Blackwell, 2002.

Brown, N., B. Rappert, and A. Webster. *Contested Futures: A Sociology of Prospective Techno-Science*. Aldershot: Ashgate, 2000.

Cochrane, P. *Tips for Time Travellers: Visionary Insights into New Technology, Life and the Future by One of the World's Leading Technology Prophets*. London: Orion Business Books, 1997.

Dourish, P., and G. Bell. "Yesterday's Tomorrows: Notes on Ubiquitous Computing's Dominant Vision." *Personal and Ubiquitous Computing*. no. 2 (2007): 133–143.

Ee Sze, T. "No 4G." *Computerworld*. 12, no. 1 (2005): 7–20. Accessed October 5, 2005. 2005)http://computerworld.com.sg/ShowPage.aspx?pagetype=2&articleid =2720&pubid=3&issueid=67

Ellis, W. "Flying Frogs and Crashed Rocketships." Warren Ellis Dot Com. 2006. Accessed December 30, 2006. www.warrenellis.com/?p=3442

Featherstone, M., and R. Burrows. *Cyberspace, Cyberbodies, Cyberpunk: Cultures of Technologies Embodiment*. London: Routledge, 1995.

Finn, C. *Artifacts: An Archaeologist's Year in Silicon Valley*. Cambridge, MA: MIT Press, 2001.

Gabrys, J. "Atmospheres of Communication." In *The Wireless Spectrum: The Politics, Practices and Poetics of Mobile Media*, edited by B. Crow, M. Longford and K. Sawchuk, 46–60. Toronto: University of Toronto Press, 2010.

Gupta A., and J. Ferguson. *Anthropological Locations: Boundaries and Grounds of a Field Science*. Berkeley: University of California Press, 1997.

Haraway, D. "Situated Knowledges: The Science Question in Feminism and the Privilege of Partial Perspective." In *Simians, Cyborgs and Women: The Re-Invention of Nature*, 183–201. London: Free Association Books, 1991.

Haraway, D. *A Kinship of Feminist Figurations. The Haraway Reader*. London: Routledge, 2004.

Herzog, L. "Iwatch to Save Apple?" May 16, 2013. Accessed June 22, 2013. www. bloomberg.com/news/2013-05-15/iwatch-to-save-apple-.html

Horst, H., and D. Miller. *The Cell Phone: An Anthropology of Communication*. Oxford: Berg, 2006.

Ingold, T. "The Temporality of the Landscape.", In *The Perception of the Environment: Essays in Livelihood, Dwelling and Skill*, 189–208. London: Routledge, 2000.

Ingold, T., and E. Hallam (eds.). *Creativity and Cultural Improvisation*. Asa Monograph. Oxford: Berg, 2007.

Jensen, C. B. "An Experiment in Performative History: Electronic Patient Records as a Future-Generating Device." *Social Studies of Science*. no. 2 (2005): 241–267.

Kaku, M. *Visions: How Science Will Revolutionize the 21st Century and Beyond*. Oxford: Oxford University Press, 1998.

Kirby, D. "The Future Is Now: Diegetic Prototypes and the Role of Popular Films in Generating Real-World Technological Development." *Social Studies of Science*. no. 1 (2010): 41–70.

Latour, B. *Science in Action: How to Follow Scientists and Engineers through Society*. Cambridge, MA: Harvard University Press, 1987.

Latour, B. *Aramis or the Love of Technology*, trans. C. Porter. London: Harvard University Press, 1996.

Law, J. *A Sociology of Monsters: Essays on Power, Technology and Domination*. London: Routledge, 1991.

Law, Jo. "Notes on the Theory of the Actor-Network: Ordering, Strategy and Heterogeneity." *Systems Practice*. no. 5 (1992): 379–393.

Le Guin, U. K. *Dancing at the Edge of World: Thoughts on Words, Women, Places*. New York: Grove Press, 1989.

Malone, M. "Forget Moore's Law." *Red Herring* (February 2003). Accessed February 15 2003. http://web.archive.rg/web/20030315195403/http://www.redherring.com/mag/ issue122/5945.html

Marcus, G. *Technoscientific Imaginaries: Conversations, Profiles, and Memoirs.* London/Chicago: University of Chicago Press, 1986.

Mattelart, A. "Mapping Modernity: Utopia and Communication Networks." In *Mappings*, edited by D. Cosgrove, 179–203. London, Reaktion Books, 1999.

Moore, G. "Cramming More Components onto Integrated Circuits." *Electronics.* no. 8 (1965): 114–117.

Nicholson, J. "Calling Dick Tracy! Or, Cellphone Use, Progress, and a Racial Paradigm." *Canadian Journal of Communication.* no. 3 (2008): 379–404.

Nowotny, H. *Insatiable Curiosity: Innovation in a Fragile Future.* London and Cambridge, MA: MIT Press, 2008.

Patel, N. *Pebble Smartwatch Review: Can Pebble Make the First Smartwatch for Regular People?* 2013. Accessed February 1 2014. www.theverge.com/2013/1/28/3924904/pebble-smartwatch-review

Pollard, J. "The Art of Decay and the Transformation of Substance." In *Substance, Memory, Display: Archaeology and Art*, edited by C. Renfrew, C. Gosden and E. DeMarrais, 47–62. Cambridge: McDonald Institute Monograph, 2004.

Porter, T. *Trust in Numbers: The Pursuit of Objectivity in Science and Public Life.* Princeton: Princeton University Press, 1995.

Rosenberg, D and S. Harding (eds.). *Histories of the Future.* Durham: Duke University Press, 2005.

Sawchuck, K. "Radio Hats, Wireless Rats, and Flying Families." In *The Wireless Spectrum: The Politics, Practices, and Poetics of Mobile Media*, edited by B. Crow, M. Longford and K. Sawchuk, 29–45. Toronto: University of Toronto Press, 2010.

Schiffer, M. *The Portable Radio in American Life.* Tuscon/London: University of Arizona Press, 1991.

Star, S. "The Ethnography of Infrastructures." *American Behavioral Scientists.* no. 3 (1999): 377–391.

Stone, A. *The War of Desire and Technology at the Close of the Mechanical Age.* Cambridge, MA: MIT Press, 1996.

Strathern, M. *Partial Connections, Association for Social Anthropology of Oceania Special Publications Series.* Lanham: Rowman and Littlefield/AltaMira, 1991.

Suchman, L. "Located Accountabilities in Technology Production." *Scandanavian Journal of Information Systems.* no. 2 (2002): 91–105.

Suchman, L. "Affliative Objects." *Organization.* no. 3 (2005): 379–399.

Suchman, L. *Human-Machine Reconfigurations: Plans and Situated Actions* (2nd expanded ed.). New York/Cambridge, UK: Cambridge University Press, 2007.

Turnbull, D. "Performance and Narrative, Bodies and Movement in the Construction of Places and Objects, Spaces and Knowledges: The Case of the Maltese Megaliths." *Theory, Culture and Society.* no. 5/6 (2002): 125–143.

Watts, L. "Designing the Future: Fables from the Mobile Telecoms Industry." In *Thumb Culture: The Meaning of Mobile Phones for Society*, edited by P. Glotz, S. Bertschi and C. Locke, 225–234. Piscataway: Transcript Verlag/Transaction, 2005.

Watts, L. "The Future Is Boring: Stories from the Landscapes of the Mobile Telecoms Industry." *21st Century Society: Journal of the Academy of Social Sciences.* no. 2 (2008): 187–198.

Watts, L. "Sand14: Reconstructing the Future of the Mobile Telecoms Industry." *The Fibreculture Journal.* no. 139 (2012): 33–57.

Part III
Living Mobile Lives

9 The Mobile Phone (and Texts) as a Taken-for-Granted Mediation

Rich Ling

Some technologies become so woven into our daily experience that they are taken for granted. In many countries, the mobile phone is beginning to have this status. In 2000, about one person in ten had a mobile phone, but by 2014 there were almost as many mobile phone subscriptions as there were people in the world. Because of dual subscription phones and other doubling up of subscriptions this translated into about 50 percent of the world's population being active users.[1] About a third of the world's population has access to the Internet, but three of four people who have Internet access can access it via a mobile phone.

The technology has established itself in a niche in our daily lives. It makes coordination of daily affairs more convenient, it allows us to stay in touch with others, and we find a sense of safety and security in having a mobile phone. Still, ownership and use of a mobile phone is not like most other forms of personal consumption because wireless phones are generally part of a socially and materially networked system. The history of the fax machine is instructive here. The fax machine started out as being a link between two specific individuals, but as it reached a critical mass, users in some social communities started to assume that everyone was available for faxing.

Having a phone facilitates many kinds of interpersonal interactions. It allows us to coordinate mundane affairs ("Hi, the bus will be at the station in about twenty minutes, can you pick me up") and it allows us to work out the small interactions of daily life ("I am a few minutes late because of traffic; I will get there as soon as I can. Thanks for telling me that you moved into meeting room three"). This nuanced micro-coordination is probably one of the most profound functions of the mobile phone (Ling 2012). It allows efficiency in planning and executing our interactions that was not possible with traditional landline telephony. In addition to being useful for instrumental interaction, it also enriches and intensifies social interaction. We can announce the small victories and setbacks to our social circle almost as they are happening.

Critical mass is not a specific number or a physical "fleet" of devices; it is a collective sense that a technology or a system is ubiquitous. Once this attitude has become generally accepted in a group, if our intended interlocutors, for whatever reason, do not have a mobile phone, it becomes a problem. That is, we have evolved a type of planning and interaction that assumes perpetual access to one another. The functionality of the mobile phone, in both the form of voice interaction but also to a surprising degree via texting—first in the form of Short Message System messages (SMS) and now increasingly in the form of apps such as Viber, Whatsapp (bought by Facebook in early 2014), iMessage, etc.,—is being woven into our daily interactions. To be sure, there are alternative ways of conducting coordination and social interaction, yet there is an increasing reliance on mobile mediated coordination. To the degree that it becomes "the" accepted form of interaction, it replaces other systems.

As this develops, the mediation form attains a set of courtesies and manners. We begin to understand the form of an appropriate and an inappropriate text message. We gain a sense of how long it can go before we answer an SMS from an acquaintance, a friend and a spouse. We have a sense that it is appropriate to be available and we scorn others who are less responsible. That is, we develop an ideology that supports the use of mobile telephony.

The sum of this diffusion, entrenchment and sense of correctness is a logic of reflexive expectations. We begin to assume that others will expect a certain type of use from us, and we expect the same from them. There is clearly slack in the particular forms and uses, but so long as we are within certain boundaries, we do not strain the boundaries of other's expectations.

It is important to remember that "taken for grantedness" is not necessarily a social universal. Obviously, before a technology can be seen as taken for granted, it has to become diffused into society. That is, it has to be used (not just purchased or obtained) by enough people in the group so that it is seen as a normal part of daily interaction. It is worth noting technologies can become taken for granted within smaller subgroups. It might be that a particular clique adopts Foursquare or perhaps MXit (Walton and Donner 2011) as their method of interacting with one another. It may well be that in some groups; members absolutely have to be available via Facebook, where in others it is MySpace, World of Warcraft, via the mobile phone, or via the landline telephone. This can vary from group to group. It can also vary from culture to culture. The point is that the item needs to be, to some degree, diffused within that particular group before it can gain the status of being taken for granted. If it is only our nearest ten friends who use a particular form of mediation, we cannot extend that expectation to people who are outside of that sphere. In this way, those systems that extend across many groups (for example texting) have the advantage of facilitating interaction with both the immediate social sphere and with others who are more irregular members of the social circle.

It is interesting to consider the degree to which the mobile phone has become taken for granted; the degree to which it is engrained into our mutual expectations; the degree to which it has established its own rationality. But it is not the only technology that performs this way. There are other socio-technological systems that function in this way. Communication technologies have a particular potential to become taken for granted. If they can garner use among enough people, they generate network externalities. This is the so-called Metcalfe Law (Metcalfe 1995). Basically, as a communication technology garners users, the total value of the medium increases as each additional user joins. Metcalfe postulated that there was an exponential increase, but this is being overly optimistic. The point remains, however, that as the technology diffuses in to society, the decision to use it is not so much an individual as a social decision.

ELEMENTS OF BECOMING A TAKEN-FOR-GRANTED TECHNOLOGY

There are several dimensions that determine whether a technology has become taken for granted. The acid test is whether we expect it of one another. If we feel that a technology or a system is expected of us and that we expect it of others in order to be counted as a competent member of a group or a society, then it has perhaps passed the acid test. If it is a problem for others when someone does use a technology or integrate themselves into a system, then the technology has gained a certain imperative.

Elements that constitute these reflexive expectations include (1) the diffusion of the technology or system; (2) our acceptance or legitimation of the technology; (3) perhaps most importantly, the degree to which we expect that others also use the technology and they expect it of us; and finally (4) the degree to which a particular approach or solution becomes dominant and indeed changes the social ecology.

Diffusion and the Perception of Critical Mass

It is clear that in order for mobile telephony, or any other technical system, to be considered as taken for granted there needs to be certain dissemination in a group or society. There are several approaches to understanding how a technology diffuses into society. Rogers's model of diffusion (1995) is one of these. It traces how different types of consumers adopt an item. There are the innovators who eagerly cast themselves over the new item, the somewhat less enthusiastic early adopters and so on. Eventually we come to the rather dull late adopters who seemingly only adopt an item after they are forced to at the point of social exclusion. Rogers's adoption curve is a slightly modified Gauss curve with the boundaries between groups generally marked by standard deviations. If we set aside this rather mechanical use of statistics for a moment, Rogers's point is that there are people who will adopt anything and who serve as a type of vanguard to try out the different items that are pushed out by the production system. The response of the first group is then filtered by what Rogers calls early adopters, who have perhaps a somewhat more sober stance. Adoption among the following groups follows a similar path, namely that their consider the experience of the previous groups before acceptance.

Rogers can be criticized for only looking at the actual purchase of an item. There is the feedback of the previous consumers when subsequent consumers make their purchase decision. However, there is an overall focus on commercial consumption. Indeed, part of Rogers's success is that he appeals to marketing managers in industry.

An alternative approach is that of domestication (Haddon 2003; Silverstone and Haddon 1996; Silverstone and Hirsch 1992). Consumption in the context of domestication is not simply a purchase decision. Rather it is the process through which an individual discovers an item, begins to think about how it might fit into his or her life, makes the decision to purchase or obtain the item, and then goes through the actual process of working the item into his or her daily routines. There is also a final stage in the domestication approach in that the individual is judged by others vis-à-vis the things that the individual consumes. Thus, the domestication approach does not enter into the examination of who is likely to be the first or the last to buy an item as with the work of Rogers. However, it considers consumption in a broader, and in a particularly noncommercial, way.

The final stage of the domestication approach opens an interesting notion. It suggests that after we have decided to bring an item (a particular type of watch, an iPhone or a certain type of sweater) into our lives and after we begin to use it, others are free to judge our sense of taste or adventurousness. Domestication does not, however, take the step of seeing the technology as becoming taken for granted. It considers some social dynamics. However, it does not look at the broader diffusion of the item, they way that mass adoption changes the social ecology, the development of legitimations or the idea of reflexive expectations.

An important issue in diffusion of mediation technology is that there must be the perception of a critical mass (Ling and Canright 2013). We need to think that there are enough others who are also consumers so that the device will be useful. Starting at the most basic level, it is absurd to be the only person in a country with a fax or a telephone. There would be no one else with whom we could correspond or talk. Beyond this, there needs to be a certain number of persons available in order for the device to have any utility. As diffusion becomes more general, the last holdouts may even be subject to the urgings of established users.

The establishment of these user networks is one of the decisive issues for communication technologies. Early landline telephone users often bought two telephones, i.e., one for the house and one for the office, or perhaps one for the sales office and one for the factory. The same pattern was seen among early fax users. As additional people bought landline phones or fax machines there was not, at the start, much utility to the dyad of users. These lonely users were entrenched in their own use pattern and perhaps unaware that others were also fax users. They were perhaps satisfied with the ability to fax orders and reports from office to factory. Indeed there were not many users. Quite often, at this point the devices do not enjoy scales of production and they are absurdly expensive. For example, Economides and Himmelberg (1995) report that fax machines cost as much as $2,000 in the mid 1980s when they first began to be popular. Eventually, however, users also began to understand that there were other independent dyads that had the same functionality. Thus, rather than only using their "paired" devices, they discovered that it was also possible to fax to other locations. At first this was a (well-heeled) elite, and perhaps a guarded province. However, as often happens with consumer electronics, the price of the devices fell. By the early 1990s the price had fallen to a tenth of the earlier price. Many people started using the system and at some point, it became an expectation. As additional people adopted the technology, it increased the utility of the device for existing users.[2] Indeed this is the important point. When a form of mediation becomes common enough that we can expect it of one another, it has reached a critical mass. I will pick up on this thread in the section below.

The diffusion of mobile phones has a somewhat different trajectory. In the developed world, the mobile phone benefited from being a new dimension to the preexisting landline system. It often gained its first foothold in a country as a device to be used by traveling salespeople (Johannesen 1981) who were not often at a fixed location. It allowed them to coordinate their work and to be in touch with clients and their superiors (Ling 2012). From this beginning it spread to other groups, notably teens and somewhat later to the populations of developing countries (Ling and Donner 2009).

There are dynamics associated with the evolution of technologies that have a critical mass. There are difficulties associated with start-up, i.e., it is difficult to establish the original pool of users. Interactive media have an advantage in that they can often play on the dynamics of social networks in the start-up phase. However, once a system has established itself, it is less vulnerable to being supplanted by alternatives since that would involve altering the practice of the entire social network. The failed substitution of Google Wave for email illustrated this. Email, for all its limitations, is a well established and mutually recognized form of interaction. If one individual or even a small group of pioneers tries to make the transition to a substitute, they will not necessarily be able to count on all their interlocutors following suit. Thus, once a critical mass is established, there is also certain stability in the practice.

There are alternatives that do supplant previous systems. The transition from postal mail to the fax and eventually the transition to scanned documents as attachments to email illustrate this. In order for there to be a transition, however, users have to either convince or be willing to be convinced by their interlocutors of the advantages of the new system. There needs to be some compassionate reason for the first people to make the transition and they need to be willing to work through the frustrations of configuring poorly worked-out technologies. As time goes on, as the system becomes better calibrated, and as there are fewer and fewer who do not use the technology, the final holdouts face increasing insistence that they make the transition.

Legitimation

The previous discussion sounds rather brutal. As a mediation technology gains critical mass, it becomes a juggernaut that rampages through society and forces people to adopt. It is important to note that we develop legitimations and justifications to buffer the adoption process. These justifications help us to explain our use of the device or service and they help us to fit it into the broader values of the society (Berger and Luckmann 1967). They can provide the individual with all the good reasons why they should adopt. In the case of the fax machine it was the speed with which important documents could be sent. In the case of the mobile phone, people often speak of the safety and the ability to reach people regardless of where they may be. Legitimation systems can also be a bulwark against adoption. We can see in the case of mobile telephony that although there are many justifications against adoption (Ling 2004), there are seemingly more that support adoption.

When a new technology appears it is often seen as inconsequential or perhaps the hobby of those who are particularly inclined. As it becomes more engrained in society and as we need to decide whether or not we will make the transition from being a nonuser to being a user, we muster all the reasons for or against the decision. We might be fearful of the complexity and we might think ourselves as being comfortable with the existing system. On the other hand, we might be beguiled by the possibilities with the alternative or we might want to show others that we are abreast of new ideas.

As technologies diffuse, these lines of argument are built up by some to defend the existing system and by others to advocate its replacement. It is in these interactions that we develop the legitimation vocabulary for our perspectives.

Reflexive Expectations

An important point in the cycle of systemic adoption is when we take it for granted that that others in a particular group are users; when we are more surprised by finding people who are not users than we are surprised by finding those who are. In other words, the use of the technology has become routine. It is what we expect. Upon reaching criticality, use of mediation technology becomes a type of public good that is independent of individuals. Universal access (within the context of whatever universe that is being considered) is a type of commons from which all can benefit. Interactive communication services embody reciprocal interdependence. The ability to quickly get information from another person, or to provide it upon their request, is a latent potential in the system (Markus 1987).

As with any other public good there are common benefits, and there are individual responsibilities. In the case of mobile telephony, the advantage to the individual is that they can send and receive texts that help to guide them to meetings and inform them as to the status of different friends. However, the individual also has to maintain the terminal. They have to pay for the subscription and make sure that the terminal is charged, etc.

Change the Social Ecology

An important element that characterizes taken-for-granted technologies is that as they become entrenched, they make alternatives obsolete. As this happens, there is an increasing dependence on the particular item. To use the metaphor from biology, they extend the niche of the item beyond its original boundaries and, in some ways, reform the surroundings.[3]

To use the example of the fax machine, it changed the way that we think about the transmission of written material and indeed it changed the structure of the institutions associated with information transfer. Previous to the popularity of the fax machine, paper-based information needed to be sent physically. A newspaper clipping, a production report, a contract, etc. all needed to be mailed in their physical form. As the name implies, the fax allowed a facsimile of a document to be sent from one point to another. This functionality was a threat to the postal system. According to Skelton et al. (1995), the substitution of faxing for the postal service came to the UK in 1987 when there was a postal strike. The fax machine had been gaining ground at this time. However, the postal strike meant that users had an additional motivation to make the transition. For what was by then a relatively small investment, they could avoid the inconvenience of the striking postal workers. Looking at this situation somewhat more broadly, the fax machine moved into the niche that had been occupied by another actor. The combination of increasingly functional technology and the perhaps not unrelated strike, fax machines gained a foothold. After they had gained this position, it was difficult for the traditional postal system to regain its position. It is clear that the same technological jujitsu happened to the fax machine with the introduction of email and scanners.

Focusing again on the mobile phone, there are some of the same dynamics, albeit in a somewhat different costume. The mobile phone arose from the melding of the switched landline system with the broadcast radio system (Agar 2003). In effect, it gave us a private telephone "line" that employed the radio spectrum. While there is an existing landline telephony system, the fact that we are individually available (Ling and Donner 2009) on a continual basis is seen as a major advantage. It allowed us to send and receive texts and to chat regardless of location. Although it diffused as an extension of the landline system, it has expanded its position so as to replace the both home phones and phone booths. Indeed in many homes people are canceling or never subscribing to the landline telephone system (Dutwin, Keeter and Kennedy 2010). In other words, the mobile phone is changing the social ecology of mediated communication. In addition, it provides for texting (something that the landline system has not done in any large degree) and it has started to provide access to the Internet through advanced handsets. It is expanding the areas in which we can make and receive calls. At the same time the system of fixed terminals is being disassembled. The mobile phone is becoming structured into our communication praxis. To use a biological metaphor, the mobile phone has out-competed the landline telephone system and, after establishing itself in that niche, it has expanded the reach of what we can expect from telephony.

TEXTING AS A TAKEN-FOR-GRANTED TECHNOLOGY

Texting is, for many, a taken-for-granted system of mediation. It has a long since exceeded a critical mass in many groups (Ling, Bertel and Sundsøy 2012), it has become legitimated in terms of the style of use and the meaning of SMS argot (Baron 2008), we often assume that others are available via texting, they assume the same of us (Ling 2012), and it has rearranged the social ecology of communication (Licoppe 2004).

Texting (either using traditional SMS or another service such as Whatsapp, Viber, etc.) is a dominant part of mobile communication. A 2009 analysis of 18,500 anonymous users of all ages in the Telenor net in Norway shows that SMS is one of the dominant activities. When counted as individual events, SMS made up more than half of what people do on their mobile phones. At that point, Individual calls make up 41 percent of the activity for normal users and data transmission (clicking on links) makes up about 8 percent of all activity. Looking at activity in the U.S., according the Pew Internet and American life project, 65 percent of 18–29-year -olds sent SMS in 2009. In 2013 it was up to 97 percent. Between 2009 and 2011 the median number of texts/day for teens 12–17 went from fifty in 2009 to sixty (Lenhart 2012). When comparing to other mediation forms, in 2013 26 percent of this age group had ever used Snapchat and 43 percent had used Instagram (Dugan 2013).

By these measures, SMS (and increasingly text messages sent through other mediation systems such as Facebook) is a central form of communication that we use in order to maintain our social ties. This functionality is likely to also be seen in various other forms of mobile texting services. Texting has moved from being a marginal form of mediation fifteen years ago to being one of the most central forms of mediated interaction. Indeed, in 2013 there were approximately ten trillion SMSs sent and received.[4] This breaks down to about eight texts per day, for all 3.4 billion mobile phone users in the world. Clearly there are differences with regards our use of SMS (Lenhart et al. 2010; Ling et al. 2012). Figure 9.1 shows the relationship of

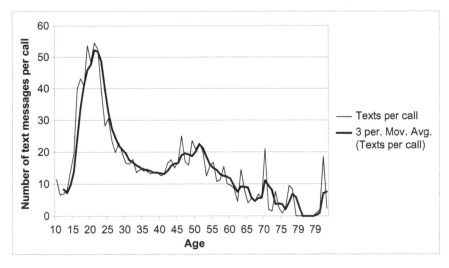

Figure 9.1 Ratio of texts to calls (three-period moving average superimposed on raw data) based on 13,931 users in the Telenor net in Norway in 2009.

texts to calls for a sample of Norwegians in 2009. Over all, the people in the sample (that included almost 14 000 persons) send almost 4.5 texts for ever call they made. It is, however the teens and young adults who are the most SMS-oriented in their communication patterns.

Texting has changed the way that we contact one another and it has changed the landscape of interpersonal interaction (Licoppe 2004). It has opened the doors for asynchronous, text-based interaction. Previous to SMS, we either had to call one another (often at landline telephones), leave notes or simply wait until we saw one another in order to exchange comments. SMS has opened a channel between individuals that did not exist before. The volume of use shows that we have taken it into our daily routines. Indeed, we arrange many facets of our lives using this form of mediation and following the broad theme of this paper, we expect it of one another. To use the formulation of James Katz, we are a problem for others when we do not use it (Ling 2012).

TAKEN-FOR-GRANTEDNESS AND SOCIAL CONSTRUCTIONISM

Suggesting that mobile telephony becomes taken for granted echoes the sociological tradition that sees social phenomena as having a type of palpable objectivity (Gilbert 1989). This perspective is somewhat different than the idea of social constructivism where it is social forces that shape technologies (Campbell and Russo 2003; MacKenzie and Wajcman 1985). On the one hand, there is the suggestion that it is the technologies that shape social practice (Schroeder 2007). There are elements of this direction that one can read into the approach taken here, namely that technological systems have a type of proscriptive tangibility. The use of the mobile phone and SMS becomes a light version of Durkheimian social facts (Durkheim 1938). That is, our collective notions of use are a social force that is difficult to avoid. They "are imbued with a compelling and coercive power by virtue of which, whether he wishes it or not, they impose themselves" (Durkheim 1938, 51).

We carry with us the sense that we are being negligent toward our social sphere when we forget our mobile phones or when we do not answer texts. We have somehow not paid due respect to our social responsibilities. There is, at some level, the sense that society is being formed as a result of the technical systems that operate in its midst. We are at the mercy of a technology and other's expectations of how we should use it. By contrast, the social constructionist perspective is interested in examining how social processes shape artifacts. It is the values of social interaction that are brought together in the forming and refining of technology. In the one case it is the technologies (or our reading of the technologies and their uses) that are setting the tempo, and in the other case it is the social processes.

It is clear that both social constructionism and technical determinism can each be used as a lens with which to help us understand the role of new technology. We do indeed appropriate and adapt technologies to our own needs and in this way we are active "cocreators" of technology. It is also true that technologies can, as Weber suggests, stiffen into what he called the iron cage (1930, 181) such that they restructure social interactions. Neither of these positions is to be denied. However, to look at the situation from a static perspective is not particularly helpful, nor is it useful to reduce the discussion to only issue of primacy.

The approach taken here encompasses both the malleability of social constructionism and the proscriptive tangibility of the more technical deterministic approaches. It proposes, however, that this is evolutionary. It is not correct to only look at one point in time but rather it is a dynamic social process. At the point of departure, many (though not all) technologies are socially pliable. This development has been seen in the rise of texting. The Internet or the PC and their many applications are also examples of this and indeed the structure of these applications means that they are perhaps irresolvably changeable. However, with time our collective acceptance of a particular technology constrains their malleability. This can, for example, be seen in the way that time and specifically the notion of "clock time" changed during its adolescence in early modern Europe. There were many different physical devices, many different ways of counting time, different notions of when a new day began, etc. Indeed medieval Europeans had hours that varied according to the seasons and there was no general sense of when one day ended and another began. Thus, the social constructionist perspective can delight in how different cultures and different societies developed timekeeping in unique ways (indeed there is still some of this as in Levine's *Geography of Time* [1998]). However, there is also a very real sense that mechanical timekeeping has become taken for granted (Zerubavel 1985). That is, it has become less malleable and we expect of one another that we both use the same timekeeping system. We have developed a more crystallized "taken for granted" relationship to timekeeping and as this has developed, it is less compliant to social shaping. There can be different advances in the devices that are used, but the system of two twelve-hour periods during the day starting at midnight is fixed. This is remarkably resistant to change.

This process of transition from malleability to crystallization is described by Berger and Luckmann in their discussion of institutionalization of social practice (1967). They describe how social artifacts (mobile phone use for example) form as a result of social interaction. We create notions of one another's actions that eventually become engrained in what they famously call the "reciprocal typification of habitualized actions."

> Institutionalization occurs whenever there is a reciprocal typification of habitualized actions by types of actors. Put differently, any such typification is an institution. What must be stressed is the reciprocity of institutional typifications and the typicality of not only the actions but also the actors in institutions. The typifications of habitualized actions that constitute institutions are always shared ones. They are available to all the members of the particular social group in question, and the institution itself typifies individual actors as well as individual actions.
>
> (Berger and Luckmann 1967, 54)

They go on to discuss how these institutions (such as the constraint to pay attention to time or to carry a mobile phone) are not developed in a power-free situation. Indeed there are forces that urge the social development of these social constructions in one way or in another. The phone companies will want to make us think about making calls, our job will want us to be available, etc. The objectification of these social constructions are the result of our collective actions and they are tempered and adjusted through a whole array of power relations. Thus there is the malleability of the constructionist perspective, the preconfiguration defined by the constraints of the technology and the suggestion that, with time, they will stiffen into social constructions that constrain and direct our behavior.

OTHER EXAMPLES OF TAKEN-FOR-GRANTED TECHNOLOGIES

Looking beyond communication technologies, it is possible to see that there are other technologies that function as taken for granted. Mechanical timekeeping and the automobile/suburban housing complex are two examples.

It is also instructive to think about well-diffused technologies that do not fulfill these criteria.

As noted, timekeeping has seen this development. The mechanical clock was first developed in the service of Benedictine monks (among others see Zerubavel 1985). As the technology became more robust and portable, mechanical timekeeping diffused through society and in many ways remade social interaction in its own image (for a somewhat strident version of this see Mumford 1934). It gained a central position in society in, for example, the UK in the late medieval period (Glennie and Thrift 2009). As it moved into society it replaced a welter of different timing systems that were, in many cases, based on bells. In a typical city previous to the diffusion of mechanical clocks, different activities were regulated through the position of the sun and quite often by signaling bells. Different institutions, each with their own schedule, would have some system of bell ringing. There were bells for religious gatherings, for civic functions (council meetings, courts, etc.) and for commercial activities. Since each institution had its own set of bells and own schedule there was a confusion of different systems. According to Dohrn-van Rossum, the "acoustic environment" could be surprisingly complex (1992). He reports that Florence had eighty different bells (or more precisely bell sets), each with their own periodicity. If a person were to go to nearby Siena, there was a different set of bells and a different local practice. Mechanical timekeeping simplified this unwieldy system with a single abstract metric that was standard for all.

As mechanical timekeeping became integrated into society, it also engendered a system that legitimized its use. While it can be seen as being stressful (Zerubavel 1985), it is also considered impolite to be late or to waste others' time (Levine 1998). The use of time is also a social institution. We expect competence in time telling from one another. If we say to one another that we will meet at two thirty, then there is the assumption that each party will understand and respect this decision. Time helps us to mediate social interaction by providing a mechanism through which we can coordinate interaction. Thus, mechanical timekeeping fulfills the notion of being a taken-for-granted technology.

The complex of automobiles and suburban living also, in some cases, fulfills the criteria of being taken for granted. It is clear that for people living in Manhattan or central Paris or London, that is cities where there are well-developed alternatives, a car is not necessary. However, in the vast suburbs of many cities such as those found in Atlanta, Sydney, the areas around Paris and London and of course Los Angeles, it is not reasonable to live without a car. The automobile, of course, has a relatively short history. It is really only in the last half of the twentieth century that there was the intense development of suburbs. In this period, however, the system of automobile-based transportation has altered the development of cities, roads, shopping and a whole array of large and small institutions. In very real ways, this development has changed the social ecology of our lives. We also have developed a range of legitimations for the automobile. We talk about how it gives us freedom of movement, the ability to seek out better opportunities, etc. Finally, it is also taken for granted. In many cities, when we make agreements to meet or to participate in activities, there is also the implicit assumption that the people who are participating will transport themselves with a car. In those cities where there is not a well-developed system of public transportation, not having a car throws the individual on the charity of their friends. This is seen, for example, when elderly people move to sheltered care facilities that are in the fringe areas of the cities. Since this change is also accompanied by losing access to a car, they become isolated from their social ties. Thus, like mechanical timekeeping, the system of suburban living supported by automobile transport has the characteristics of a taken for given system.

One final example of a technology that is widespread but does not fulfill the criteria helps to illustrate the concept. Mechanical refrigeration can be seen as perhaps one of the most significant technical developments of the last 150 years. It has facilitated the preservation of food, changed the quality of the food we eat and has also contributed to the commercialization of food production. Refrigeration in food transport and in grocery stores has changed the variety and the quality of the food we eat. Further, in many countries it is rare not to have a refrigerator in the home.

Refrigeration is also a technology to which we pay little heed. At the same time it is an essential foundation for modern life. In-home refrigeration is widely diffused, it has changed the social ecology and we have a wide variety of legitimations for its use. However, coming back to the idea of social mediation technologies, there is not a sense of reciprocity associated with mechanical refrigeration. With the mobile phone we are somehow failing our friends if we do not have one. If we do not respect time or if we are always the one who needs to be driven we also see the proscriptive tangibility of these technologies. By contrast, if we do not have a refrigerator it does not reverberate through our social circle to the same degree. Thus, it is not an example of a taken-for-granted technology to the same degree as mobile phones, clocks and cars.

CONCLUSION

It is often commented that mobile telephony and texting in its various forms has established itself in our daily lives. It is a convenient way to stay in touch with one another, to coordinate activities, to reach one another in an emergency or as a way to engage in loose talk.

The system has grown dramatically since the 1990s to the point that it is perhaps the most common type of technology available around the globe. It is possible to call a person in China, Uganda, Haiti as well as Seoul, New York, London or Kinsarvik no matter where they are at the moment. In the developed world, there is often social pressure on the laggards who do not own and use a mobile phone. Their obstinance might be an individual character trait, but it also hinders the smooth functioning of social commerce. The system of making appointments, updating one another and saying in contact has slowly become the province of the mobile phone. Our relationship to the device has made a gradual transition from being something that it "nice to have" to being an integral tie to our social network. We have come to recognize that it is more efficient to have one than not.

Along the way we have developed an ethic that we should be available to one another via the phone. Thus, we do not only have a functional relationship to the device, we have developed a legitimization structure describing its social position. Clearly there are times when we wish it was not that way, but on the whole, we have accepted its role in our lives. Indeed, we demand it of one another.

This has grown to the situation that there is a critical mass of mobile phone and there is a strong legitimation structure. Thus, we have begun to assume them of each other. We assume that we can always call one another to change the time or the place of a meeting. We can send a text to cheer up a sick friend or to ask how they did on their final exam. When they have forgotten their phone or when they are, for whatever reason, not available, the social fabric is torn in some small way. All of this suggests that the mobile phone is a taken for granted form of social mediation. Just as time and timekeeping or the automobile in suburbia is taken for granted, so is the mobile phone.

Beyond being a convenient functionality, we can, in a moral sense, demand them of one another. We, in the sense of the collective social expectations, can ask that a person have a mobile phone, just as we can ask that they pay attention to the time or that they take responsibility to transport themselves in suburbia. The weight of these social expectations comes is not the fault of any particular individual, rather it comes from the interaction of us all working out how to deal with the daily demands we face. The car, the clock and the phone have become three of the technologies that support us in the process of getting to work, agreeing on the location and time of meetings, picking up the kids at school and getting to the store before it closes. At the personal level there is an efficiency that comes from collective adoption. It is easier to coordinate; it is quicker to get in touch, etc. At the same time there are side effects. The suburbanization of society has led to pollution. Clock time leads to stress and this can in turn lead to ulcers and other physical maladies. The mobile phone can have the same impact.[5] For better or worse, that is the nature of the situation.

NOTES

1. See https://gsmaintelligence.com/.
2. This has been called Metcalfe's law, saying that the value of a network increases with the number of connections. The degree to which it increases is an issue that has been raised in this discussion (Briscoe, Odlyzko and Tilly 2006).

3. Sawhney (2005) has described this in the transition from canal transport to train transport.
4. See www.marketwired.com/press-release/-1748618.htm.
5. There is an ongoing discussion as to the effect of electromagnetic radiation. It is beyond the scope of this paper and the competence of the author to consider this issue.

SOURCES

Agar, Jon. *Constant Touch. A Global History of the Mobile Phone.* Cambridge, UK: Icon Books, 2003.
Baron, Naomi. *Always On: Language in an Online and Mobile World.* Oxford: Oxford University Press, 2008.
Berger, Peter, and Thomas Luckmann. *The Social Construction of Reality: A Treatise in the Sociology of Knowledge.* New York: Anchor, 1967.
Briscoe, Bob, Andrew Odlyzko and Benjamin Tilly. "Metcalfe's Law is Wrong—Communications Networks Increase in Value as They Add Members—But by How Much?" *IEEE Spectrum.* no. 7 (2006): 34–39.
Campbell, Scott, and Tracy Russo. "The Social Construction of Mobile Telephony: An Application of the Social Influence Model to Perceptions and Uses of Mobile Phones within Personal Communication Networks." *Communication Monographs.* no. 4 (2003): 317–334. DOI:10.1080/0363775032000179124.
Dohrn-van Rossum, Gerhard. *History of the Hour: Clocks and Modern Temporal Orders.* Chicago: University of Chicago Press, 1992.
Dugan, Maeve. "Cell Phone Activities." Pew Research Center. September 16, 2013. Accessed November 1 2013. www.pewinternet.org/files/old-media//Files/Reports/2013/PIP_Cell Phone Activities May 2013.pdf
Durkheim, Émile. *The Rules of the Sociological Method.* New York: Free Press, 1938.
Dutwin, David, Scott Keeter and Courtney Kennedy. "Bias from Wireless Substitution in Surveys of Hispanics." *Hispanic Journal of Behavioral Sciences.* no. 2 (2010): 309–328.
Economides, Nicholas, and Charles Himmelberg. "Critical Mass and Network Evolution in Telecommunications." In *Toward a Competitive Telecommunication Industry*, edited by Gerald W. Brock, 47–64. Mahwah, NJ: Laurence Erlbaum, 1995.
Gilbert, Margaret. *On Social Facts.* London: Routledge, 1989.
Glennie, Paul, and Nigel Thrift. *Shaping the Day: A History of Timekeeping in England and Wales 1300—1800.* Oxford: Oxford University Press, 2009.
Haddon, Leslie. "Domestication and Mobile Telephony." In *Machines That Become Us*, edited by James E. Katz, 43–56. New Brunswick, NJ: Transaction, 2003.
Johannesen, S. *Sammendrag av markedsundersøkelser gjennomført for televerket i tiden 1966—1981.* Kjeller: Televerkets Forskninginstitutt, 1981.
Lenhart, Amanda. "Teens, Smartphones & Texting." Pew Internet & American Life Project. March 19, 2012. Accessed February 3 2014. www.pewinternet.com/~/media/Files/Reports/2012/PIP_Teens_Smartphones_and _Texting.pdf
Lenhart, Amanda, Rich Ling, Scott Campbell and Kristen Purcell. "Teens and Mobile Phones." Pew Research Center. 2010. Accessed November 15 2012. http://www.pewinternet.org/2010/04/20/teens-and-mobile-phones/
Levine, Robert. *A Geography of Time: The Temporal Misadventures of a Social Psychologist, or How Every Culture Keeps Time Just a Little Bit Differently.* New York: Basic, 1998.

Licoppe, Christian. "Connected Presence: The Emergence of a New Repertoire for Managing Social Relationships in a Changing Communications Technoscape." *Environment and Planning D: Society and Space*. no. 1 (2004): 135–156.

Ling, Rich. *The Mobile Connection: The Cell Phone's Impact on Society*. San Francisco: Morgan Kaufman, 2004.

Ling, Rich. *Taken for Grantedness: The Embedding of Mobile Communication into Society*. Cambridge, MA: MIT Press, 2012.

Ling, Rich, and Geoff Canright. "Perceived Critical Adoption Transitions and Technologies of Social Mediation." Paper presented at The Cell and Self Conference in Ann Arbor, Michigan, 2013.

Ling, Rich, and Jonathan Donner. *Mobile Communication*. London: Polity, 2009.

Ling, Rich, Troels Fibæk Bertel and Pål Roe Sundsøy. "The Socio-Demographics of Texting: An Analysis of Traffic Data." *New Media & Society*. no. 2 (2012): 280–297.

MacKenzie, Donald, and Judy Wajcman. *The Social Shaping of Technology: How the Refrigerator Got Its Hum*. Milton Keynes: Open University Press, 1985.

Markus, Lynne. "Toward a "Critical Mass" Theory of interactive Media: Universal Access, Interdependence and Diffusion." *Communication Research*. no. 5 (1987): 491–511.

Metcalfe, R. "Metcalfe's Law: A Network Becomes More Valuable as It Reaches More Users." *Infoworld*. no. 40 (1995): 53–54.

Mumford, Lewis. *Technics and Civilization*. San Diego: Harvest, 1934.

Rogers, Everett. *Diffusion of Innovations*. New York: The Free Press, 1995.

Sawhney, Harmeet. "Wi-Fi Networks and the Reorganization of Wireline–Wireless Relationship." In *Mobile Communications: Re-Negotiation of the Social Sphere*, edited by Ling and Pedersen, 45–61. London: Springer, 2005.

Schroeder, Ralph. *Rethinking Science, Technology and Social Change*. Stanford: Stanford University Press, 2007.

Silverstone, Roger, and L. Haddon. "Design and the Domestication of Information and Communication Technologies: Technical Change and Everyday Life." In *Communication by Design: The Politics of Information and Communication Technologies*, edited by Silverstone and Mansell, 44–74. Oxford: Oxford University Press, 1996.

Silverstone, Roger, and Eric Hirsch. *Consuming Technologies*. London: Routledge, 1992.

Skelton, C., T. Lynch, C. Donaldson and M. Lyons. *Modelling Product Life Cycles from Customer Choice*. Ipswich: British Telecommunications Laboratories, 1995.

Walton, M., and J. Donner. "Read-Write-Erase: Mobile-mediated Publics in South Africa's 2009 Elections." In *Mobile Communication: Dimensions of Social Policy*, edited by James E. Katz, 117–132. New Brunswick, NJ: Transaction Publishers, 2011.

Weber, Max. *The Protestant Ethic and the Spirit of Capitalism*. London: Routledge, 1930.

Zerubavel, Eviatar. *Hidden Rhythms: Schedules and Calendars in Social Life*. Berkeley: University of California, 1985.

10 New and Old, Young and Old
Aging the Mobile Imaginary

Barbara Crow and Kim Sawchuk

The global population is an aging population. According to World Health Organization (WHO) projections, the proportion of people who are sixty and over is growing faster than any other age group. Between 1970 and 2025, a growth in older persons of some 694 million, or 223 percent, is expected, with 80 percent living in developing countries. At the same time another transition is occurring: there is a marked increase in the adoption of the cellular telephone and networked services worldwide. It is estimated that subscriptions will reach five billion (83%) by the end of 2010 (Whitney 2010). These rates of adoption are not evenly spread throughout the world or within populations. For example, in 2008, 87 percent of the European Union population between sixteen and seventy-four years old declared themselves to be mobile owners or users. Within this number, only 62 percent of seniors aged sixty-five to seventy-four years old declared themselves to be users of this increasingly ubiquitous means of communicating, although this number is rapidly changing as the population, habituated to mobile phone technologies and services, ages.

Despite their demographic importance, this population is systematically left out of both statistical studies generated by the wireless industry and critical media studies on everyday media uses and practices. In our ongoing research on the interlocking intersection of a discursive assemblage of mobility, aging and neoliberalism, one of the notable trends in the media coverage, marketing and the academic scholarship on mobile telephony is a focus on the practices of the young, the promise of the new and the problem of the old (Katz and Aakhus 2002; Ling 2004). Moreover, while some work focuses on seniors and technologies, an overwhelming number of studies do not distinguish between older users, who are simply categorized as 55+1. Unlike the gradation accorded to younger users, whose age-based capacities are demarcated in five year increments from 14 to 17, 18 to 24, 25 to 34, 35 to 44, 45 to 54– followed by 55+;[1] the older you become the more homogeneous the data collection on media use practices. In this chapter, we examine the perspectives and use practices of those who are sixty-five and older. We do so to explore the tangled tensions and animated intersections between aging, mobile technologies and the broader issue of mobility, a tracing of what we might understand as a 'mobile assemblage.'

The term *mobile assemblage* is adapted from Bruno Latour's notion of the "the social" as an assemblage of actions, rather than as a noun or fixed entity. A mobile assemblage is not just indicative of the presence of an array of mobile devices, from PDAs to wireless laptops, but an assemblage of practices, sets of relations and fluid associations between both human and nonhuman actors. As Latour suggests, to understand the contours, detours and sedimentations of power in this assemblage "you have to follow the actors themselves, that is try to catch up with their often wild innovations in order to learn from them what the collective existence has become in their hands, which methods they have elaborated to make it fit together, which accounts could best define the new associations that they have been forced to establish" (Latour 2007, 12).

And so we present six stories from a much larger corpus of interview material conducted over a three-year period of research with an array of participants who are between the ages of sixty-five and eighty-four (Crow and Sawchuk, 2009–2011). These stories are both unique and yet oddly representative, as they capture the movement-experiences of research participants as they negotiate both urban and rural spaces, contemporary communications technologies, and the physical and corporeal challenges that come with aging. They are documents of some of the many ways that these participants have fashioned for themselves what the gerontology literature terms as "active ageing" (WHO 2002, 12).

This neologism has been promoted within the field of aging studies within the past decade. As an international policy paper on aging states, "Active ageing allows people to realize their potential for physical, social, and mental well-being throughout the life course and to participate in society, while providing them with adequate protection, security and care when they need" (WHO 2002, 12). The report goes on to define activity as "continuing participation in social, economic, cultural, spiritual and civic affairs, not just the ability to be physically active or to participate in the labour force" (WHO 2002, 12).

The active-aging agenda is intended to rectify cultural perceptions of aging as simply a period of decline associated with a lessening of our physical and mental capacities. The active-aging agenda is a laudable attempt to address myths about aging, yet strangely absent from many of the reports being produced by various governmental agencies is the potential role of communications and culture or media technologies within this agenda. With a tendency to provide composite portraits and statistical evidence, what is also notably absent are the textured voices of seniors themselves.[2]

MOBILITY AND AGING: THE SILENCING OF SENIORS

In her monumental study of aging, Simone de Beauvoir (1974, 10) noted the elderly have been "silenced" in many a society and are typically left out of the cultural landscape: "there are books, periodicals, entertainments, radio and television programmes for children and young people: for the old there are none." While there is nonetheless an increasing attention to what has been termed the new seniors' market, within media studies, particularly those that deal with digital media, one notes a preponderance of attention on those who are young, and scant attention to those who are "older." It is worth asking if there exists an implicit discursive identification and slippage of "new media" with the "young" in our understanding of those who use digital media, and how this elision might be affecting the shaping of the present mobile assemblage.

As Lisa Gitelman (2008) argues, "new" forms of media are represented and marketed to us as better, faster and more convenient, and almost always in an ahistorical context. Some of the tacit consequences of the emergence of this discourse in a neoliberal conjuncture is a trend to the increasing privatization of public life, an emphasis on heightened consumerism, the growth of mobile and precarious labor, and an economics of perpetual growth. Within this agenda, to be dated is to be fated.

Marketing discourses are imbedded with a language that equates the "life cycles" of products with the human life cycle. This life cycle is commonly described as a journey from the immobility of infancy, upward to the peak of fully mobile, and masculine, maturity, then downward to elderly immobility (Sawchuk 1995). In the press and policy documents, aging is often represented with dire warnings that we "must deal with an aging population" and of an impending "grey tsunami" that is about to drown the nation (Mitchell 2010). Aging is positioned in pejorative terms as largely an economic condition that will bring hardship to the budget of the nation and to the next generation of taxpayers: the old are figured as a burden instigating a fear not only of the aged, but of aging itself.

In academic scholarship on mobile technologies, most research focuses on what is next, not what "has been." In this epistemology of the new, researchers tend to examine how youth adopt mobile technologies and the ways in which they are changing communication practices (Ito, Okabe and Matsuda 2005). Those who are 55+ are part of a category, often viewed pejoratively in studies of technological diffusion, as nonadopters or as "basicans" who do not take, in the words of one researcher "full advantage" of the phone's multi-functionality (Lee 2008).

One of the earliest works on the elderly and cell phones is arguably Rich Ling's (2008) overview of SMS texting in Norway. As Ling writes, the "difference" between the old and the young "is glaring." While youth have a "passion" for the device, the elderly do not have the same "need to keep track of buckets of friends." In Ling's terms, teens "breathe life" into their mobile phones; however, the same assumption cannot be made in reference to the elderly. He asks:

> "Are the elderly victims of poor design and exclusion or are they agents in their own isolation? Should the policy be to include the elderly in every possible digital development? Is their quasi self-chosen aloofness a result of poor design? Will the gap eventually disappear?" (Ling 2008, 338).

Ling's questions are pertinent, and vitally important. However, it is our contention that a deeper epistemological issue needs attention, and here we ask the question: is there a proclivity in media studies to equate the most active user with the ideal mobile subject, a perspective that orients the research agenda toward an assumption that *more* is better? How does this cohort of users, typically left out of studies of new media practices, wrestle with life in an era where the adoption of new forms of telecommunications is assumed as not only to be desirable, but is tacitly promoted as a normative value? When is skepticism toward the promise of technology interpreted—and dismissed—as an old-fashioned, stubbornly individualistic "refusal to adopt?" In what ways do researchers within the field of mobility studies themselves adopt this position?

For example, in his most recent collection on mobile communications technologies, James Katz (2008, 3) argues:

> More generally, refusal to adopt often stems from individual (and self-described) archaic views of manners, standards, and good public behavior. The decision is indicative of a life-philosophy that asserts that one must live by one's own rules, regardless of what others think.

Katz, to be fair, lists a whole host of reasons for this refusal, including being out of reach to others or avoiding situations that you might not like. However, Katz's comments reveal a more generalized attitude within media studies equating adoption with individual will and the assumption that all people—everywhere and at all times—have the choice, the time, the money and/or the physical capacity to engage within this broader landscape of mobile telecommunications. As we have pointed out elsewhere the division between user and nonuser is less a binary than a continuum, and there are also often economic and infrastructural reasons to limit one's use of these devices.

To return to de Beauvoir (1974, 10), economists and legislators endorse this convenient fallacy such that the aged are an entirely different species apart from the rest of humanity—"when they deplore the burden that the 'non-active' lay upon the shoulders of the active population". De Beauvoir's reflections on aging tease out the paradoxical values of consumer society and bourgeois culture that simultaneously repudiates aging and requires the aged to "be a standing example of all of the virtues" (1974, 11). The problem of aging, she writes, is "passed over in silence," a silence which, she says, "must be shattered." She calls upon her "readers to help me in so doing" (1974, 16).

MOBILE PRACTICES AS "STORIES-SO-FAR"

In response to de Beauvoir's still relevant plea to shatter the silence associated with aging, we present this set of stories-so-far drawn from interview transcripts with over two hundred people aged sixty-five plus, living in Canada. To date, we have held formal group discussions with over two hundred people who are sixty-four and over, accepting invitations into their community centers, legions, church halls, and homes. We have engaged in a small number of early one-on-one interviews and countless informal conversations with retired individuals in shops, on the street and in cars on the subject of aging and technological practices. We have received unsolicited emails from retired people, who have offered encouragement and their own testimonials upon hearing of our project. While technically, only the interviews have been approved by our university research ethics committee, these conversations constitute valuable source materials for understanding the digital desires and frustrations of this cohort of users who have not merely answered our list of questions, but have made many a detour in conversations and shared their stories and anecdotes.

The anecdote has long been used as a method of philosophical reflection and social criticism within feminist cultural studies. Its practitioners include Meaghan Morris (1988; 2006), Elspeth Probyn (2004) and Jane Gallop (2002), who have advocated for the use the anecdotal fragment to highlight how a singular story may be articulated to a larger set of social and cultural practices. As Meaghan Morris has written,

> They are oriented futuristically towards the construction of a precise, local, and social discursive context, of which the anecdote then functions as a mise en abyme. That is to say, anecdotes for me are not expressions of personal experience, but allegorical expositions of a model of the way the world can be said to be working.
>
> (Morris 1988, 7)

Yet while these stories are larger fragments of a larger lexicon on aging, they are allegorical expositions of how our subjects perceive the world to be working, experience the world working around them and contend with the conditions of their aging existence.

What inspires our use of these vignettes is the concept of "stories-so-far," a term introduced by feminist geographer Doreen Massey. As Massey explains, space is the agglomeration of collections of stories that occur within broader relations of power, "the power-geometries of space" that give places their character.

If space is rather simultaneity of stories-so-far, then places are collections of those stories, articulations within the wider power-geometries of space. Their character will be a product of these intersections within that wider setting, and of what is made of them, and, too, of the non-meetings-up, the disconnections and the relations not established, the exclusions. All of this contributes to the specificity of place.

(Massey 2005, 130)

Massey adds that "[t]o travel between places is to move between collections of trajectories" (2005, 130), a conceptualization of space that pays attention to movement-terms. Stories are never just there to be repeated, they are not just repertoires that help us make sense of the world, they are provisional and incomplete snatches of speech that change with every telling, and point to a notion of the subject always engaged in an act of self-transformation.

It is within this context that our "stories-so-far," told from within the mobile assemblage, profile how our interviewees talk about their experiences of mobility and movement. What does it mean not only to walk through the city with a mobile phone in hand, but with a mobile in hand as an elderly person walking, driving, scooting, taking Wheel-Trans or the bus? How does the cellular telephone fit into what we understand as a mobile assemblage, and the agenda of active aging that many of those we interviewed adhere to implicitly and explicitly? It is to these stories that we now turn.

Walking: "I'd Rather Have a Whistle"

Jacqueline,[3] eighty-four, lives alone on the sixteenth floor of an apartment in a mid-sized Prairies city. Francophone by origin, she has been twice married. From her first marriage, she has three children, seven grandchildren, and four great grandchildren. She has owned a computer since 2008, and has embraced a range of digital technologies to do research, to connect with friends and family by Skype, and to share and print photos. A security guard in a large department store during her working life, she loves to walk every day and values her independence, although driving is no longer an option. Her location provides what she needs, and she is within ambulatory range of many essentials. Across the street from her apartment is a large grocery store, where she shops, even in the frigid prairie cold when the temperatures easily dip to minus forty. In the winter, she carefully negotiates the trek from her building to the store and back across an icy and often isolated parking lot. Concerns about her safety prompted one of her grandsons to give her his old cell phone with a three-month service contract. Jacqueline took it, as she is a self-identified late-blooming nerd, but did not simply accept this as a gift as a necessity. She tested it out. Days after receiving her phone, she took her purse and the phone and did a trial run to the grocery store, at night, in the winter to test the imagined-use scenarios presented to her. "What if

I fell on the ice?" she asked, recounting this tale in an interview, "Or was attacked?" "Could I get to the phone in my bag?" She paused and laughed: "I'd rather have a whistle." She never renewed her contract.

Busing It at Night: Performing Telecommunications

Another one of our participants, Francis, came to Canada from the Philippines over fourteen years ago. Originally trained as a nurse, she now makes her living giving therapeutic massages. One of our most talkative participants, with great energy, she travels all over the city on public transit to meet her clients: some more than an hour away from her home. It was her son who suggested that a cell phone would be a way for her to keep connected to him. It also turned out to be useful for arranging appointments.

She later told us that when she traveled late at night, she would pretend she was having a conversation with someone on the phone to indicate either not to bother with her or that someone knew where she was; however, her phone was not on. She found the cell phone service too expensive. To avoid the costs, she performed telecommunications. This was a tactical practice to decrease her risk and concern about being followed or having something taken from her.

Scooting Around: Sim-Less Cell Phone

Rita is a woman in her late seventies, who lives alone in a large downtown urban center in Canada. Most of her money, from her meager pension, goes into paying for the costs of her housing. Rita has problems with physical mobility: she can no longer walk but instead relies on a motorized wheelchair, or scooter, to get out of her apartment and into the city, to shop, to go to the park or to visit friends in her neighborhood downtown community center. Rita owns a secondhand cell phone, given to her by her son, who was concerned for her safety on the busy streets. Rita does not have a service contract. She cannot afford even the ten dollars a month that some of her friends with cell phones can include in their budgets. Her rent is too high.

When asked about mobile wireless technologies, Rita tells us that she uses her *cordless* phone at home. For her it is as much a part of her repertoire of mobile technologies as any other. It stays with her so that at any time in any room she can talk with family and friends for extended periods. No extra charge, a flat rate, her landline phone is essential to her well-being and sense of connectedness. But when she is on the streets, traveling on her scooter, which she evidently loves to do, she brings her cell phone. It is with her, on her scooter at all times: "I almost got hit by a car today" she tells us. "Some idiot opened the door and almost knocked me over. Thank God I had my phone."

Without a service contract or SIM card, Rita has learned that she can still contact the police in case of an emergency. "I can't talk to friends," she says, but "If there's trouble, all I need to do is press 911 and the police are there in minutes." Rita knows. She once found a dead body in her local park and dialed the police on her SIM-free cell phone. Just as her cordless is by her side indoors, Rita never leaves the house without her cell phone. It is always charged, always on the minute she leaves, offering her the safety and security she needs at a price she can afford.

Waiting for Wheel-Trans

For many of our participants, being able to get around physically was integral to their sense of mobility and quality of life, from the rural group of quilters we interviewed who carried cell phones in their cars while they traveled from their homes in the country to the city for their guild meetings to the "snow birds" we spoke to seeking their next hotel on their long travels from Manitoba to Arizona. While these examples represent how cell phones extend their auto-mobility and connectivity, for others the cell phone acts a tool to micromanage their movements on public transit.

Carol lives in a high-rise public housing unit and needs a walker to move significant distances. Because of her assisted walking device, she has access to the public-transit pick-up and drop-off service Wheel-Trans. She uses this service to visit family and friends or to go to a doctor's appointment. In this instance, the cell phone is used as a device to extend her mobility.

The Wheel-Trans service calls when they are on their way for pick-up, providing passengers with a thirty-minute window. Upon receiving this call, Carol would head downstairs to wait. Sometimes she was held up waiting for the elevator—and she would miss the ride—as it had arrived in the first ten minutes of the thirty-minute window. When the elevator did not pose any delays, she would wait the prescribed thirty minutes. When it did not arrive, off she went back up the elevator to her apartment to call Wheel-Trans, only to find that she had missed it making the phone call! The cell phone allowed her to go downstairs and wait for bus knowing that she could be reached by the service and she could reach the service. This decreased her anxiety about whether she would miss the bus, it gave her a way to stay connected to the transit service, and ultimately, it allowed her to get to her destination.

Lost and Found: The Spousal Locator

Shopping malls and mega stores: we have all been in them. We have all lost our way. We have all lost someone who we were shopping with, someone who wanted to browse for items holding no interest for us and meandered off. He wants to look at the tools in Home Depot; she starts off in the draperies and before you know it she has wended her way to the lamp section. It's so easy to lose oneself in the maze of aisles.

Panic sets in. Your feet are tired. Walking and searching for that lost someone is no longer what it used to be for you, however. Your patience runs out. You fatigue more easily. Your bladder aches, and so you reach for your phone, you push the buttons, they answer, you save time and effort. They are relieved. A fight is averted.

For those well-off seniors we interviewed, the mobile phone offers a different kind of emergency purpose for enhancing mobility when both can afford a phone and the service contract. In this case, the phone is a finding mechanism. "It's not a phone," quips one elderly gentleman looking fondly at his wife sitting beside him on a couch in the common room of a luxurious apartment block on the Pacific West Coast: "It's a spousal locator." Everyone else sitting in the room laughs, nodding their heads. They too have used the phone in this way. Soon the room fills with dramatic stories of how to best practice spousal location. "You don't even have to answer," adds another "sometimes you just need to hear the other phone ring and follow the sound!" Not answering the phone can save money, with the mission still accomplished. None of us will ever think of the cell phone as a mobile device in quite the same way after this conversation. It is forevermore a potential spousal locator, a GPS without the need for a satellite.

Roaming: Sneaky Seniors

For those with significant retirement funds, the phone presents the ability to stay in touch. The more affluent seniors, and ones who had retired from positions with technology from their paid workplaces, understood the ways in which they could use these devices to maximize their personal efficiencies. Examples included searching for hotels when moving from destination to destination, accessing maps to provide them clear and fast directions, and calling family to let them know where they were. However, the high costs of the cell phones in the context of Canada were encountered by even the most affluent seniors in their experiences with roaming.

Despite being savvy, some were sideswiped by their high cell phones bills as the result of roaming. In one group, George recounted his strategy to decrease roaming charges:

> on my speed dial directory . . . put the one in front of everything, you won't get charged long distance for it unless you're actually truly calling long distance. So that helps too. By the way, we're all known as SS's, sneaky seniors. (laughter)

However, George did not consider himself "sneaky enough." He then went on to relay how he had bought a service to stay in contact with his journalist daughter. They purchased a cell phone package to connect to the Internet. When they returned from their three months away, they received a cell phone bill for $2,000. They immediately went to the service provider and demanded to know why they were charged this much given what they were told and what their contract said. As a result of the contract, and his persistence, the charges were dropped. However, other seniors who told us of other phone bills over $1,500 were not necessarily able to parlay their cultural capital into an effective strategy to negotiate with service providers. Many of the seniors were angered by "misleading" promises and elaborate phone contracts. They were keen to learn from others how to challenge the telecommunication providers and avoid such charges.

REFLECTIONS: STORIES-SO-FAR

Mobility, as Cresswell (2006, 22) reminds us, "is about meaning and how the resulting ideologies of mobility become implicated in the production of mobile practices." These are captured in the ways these seniors move as they age—walking, busing, driving, scooting, shopping and roaming. While not representative, these anecdotes, or "stories-so-far," typify some of the ways that seniors use and do not use mobile technologies. They reveal the significance of context in the everyday lives of seniors and the importance of teasing out and understanding both the simple and complex ways they negotiate mobility, movement and mobile technologies.

In their negotiations with individuals, friends, families, public institutions and corporations, mobility, and the cellular telephone in particular, is understood as integral to feeling safe, getting somewhere, maintaining independence and asserting agency throughout the life course. Yet, what is also revealed is the complexity of living within the current mobile assemblage, a complexity that indicates that independence and agency are always in relation to others at all times in our lives as aging subjects.

In terms of families, except for the most affluent seniors, many in this study received "hand-me-down" mobile devices from sons or daughters, although in the stories we heard it was most often sons who handed over their devices when they upgraded their own. Family members offered their used mobile devices as a way to secure their parents' safety in emergency situations. While family members are often comforted by giving their parents access to a mobile technology and emergency services, our stories also reveal that in their local contexts, like walking in the middle of winter, these devices may not be appropriate for maintaining movement. Such refusals are not about bowing out of social relations, or rejecting the phone, with aging. In the scenario "I'd rather have a whistle," Jacqueline indicates that her refusal to use the cell phone is not only about a stubborn refusal to communicate, but a desire for agency within her family nexus and a deep skepticism that technology will provide the answer to her particular set of needs and desires. Her creation and implementation of a test-case use scenario was important to her decision to determine how the device might provide her with "safety and security." Her rejection of the phone, because of her rational decision thought through step-by-step, superseded her desire to address the concerns of a worried grandson.

This anecdote points to some of the social repercussions of marketing campaigns targeted at parents and youth about the safety and security benefits of mobile phones. The term *helicopter parents*, developed to describe relations of surveillance between parents and children, could equally be applied to the relationship between middle-aged children and older parents. This corporately manufactured familial buy-in may operate as a surveillance regime, infantilizing us as we age. Mobile devices with special features to screen calls, route calls through parents' phones, or with special buttons that give instant access to family members for emergency support are indications of the construction of seniors as objects of violence and fear in need of protection as a part of the marketing discourse surrounding new technology. The mobile phone may be an apparatus to "keep them safe" or to "let someone know they are in trouble," but we also see that in these stories seniors are relaying their requirements, and their own terms, for ensuring safety and security. For some, this is not always equivalent to always being connected (Sawchuk and Crow 2010).

While receiving "hand-me-down" cell phones is a practice to make seniors feel safer and give access to emergency contacts, it is also a way to mitigate the price of participating in cell phone culture. The costs of the device and service charges are a significant concern for almost all seniors that we have interviewed, across the country and across social classes. Performing telecommunications, pretending to talk to someone on the phone while it is off, is a way to convey connection, while avoiding either the cost of use or, in other instances, the need to bother a family member, or create unnecessary worry. In the same way learning that one could use the phone, without a SIM card, was a way for another of our seniors to ensure that she was comfortable heading out into public space alone. These strategies take up the affordances of safety and security through a very precise set of limitations for using the device, but not in the manner intended by the industry. Such practices point to the tactical use practices that seniors may also share with other users, such as teenage girls. Many of our seniors passed on tips on what were the best bundling packages, which phones were easier to use and how to minimize roaming costs. The spousal locator is yet another innovative adaptation of the mobile device that allows this cohort to find and keep in touch with each other. In this case, the phone was not a technology to assuage the anxieties of their children or to ward off unwanted interlopers late at night. The mobile phone is being used as a micromanagement locating tool between husband and wife within relatively short distances within a mall.

The sneaky seniors, on the other hand, figured out ways to use the mobile device to keep in contact with their children over long distances. They actively sought out particular kinds of roaming packages and ways of using their phones to stay in communication and up-to-date on family happenings while they were away. These devices were often an extension of their physical mobility and ability to travel after retirement. Unlike those seniors who felt the device was being used to monitor them, these seniors were interested in communication with their families as they wandered the globe.

The stories indicate, as well, the range of activities and the power-geometries of class, race and gender striating the seniors' population: hardly a homogenous group. As de Beauvoir noted in 1974: "[a]lthough old age is considered a biological fate, [it] is a reality that goes beyond history, it is nevertheless true that this fate is experienced in a way that varies according to the social context" (18). As she also adds, "the individual possess a fluid, adaptable potentiality that does age, and a set, crystallized part of the mind, made up of acquired mechanisms, that does not" (de Beauvoir 1974, 40). Seniors in our discussion groups were located in public housing units and high-end retirement apartments; they were Caucasian and Black, they were widowed, married, divorced and single. Gender, race, class, age, marital status and location mattered and played a role in how they engaged and negotiated the everyday use of mobile technologies. The more affluent members of our seniors' cohort clearly had the widest range of options and use practices.

Our discussions with seniors about these matters reveal the ways in which they adapt their use of mobile technologies based on their budgets and mobility capabilities, refuse mobile technologies because they do not provide affordances to their mobility, and subvert the high costs of mobile technologies. Most of our seniors have experienced the launch, development and demise of numerous technologies, from the car, the radio, the television, the eight-track, the computer and the Internet. They have been through generations of technologies modified by the "new." Their reveal to us is how the "new" keeps them in abeyance, how the "new" allows for the further entrenchment of youth with change and innovation, and finally, how the "new" excludes other kinds of uses, practices and needs that could possibly lead to more challenging and provocative mobile technologies.

As these stories-so-far reveal, our seniors are engaged in a struggle to maintain movement and their personal mobility by using mobile phones when necessary. The seniors we have interviewed and observed over the past three years want and need mobility and adventure throughout their life course. A scooter extends their physical mobility, transportation for the disabled provides them with ways to access the communities, wireless phones allow them to keep connected from anywhere in their housing unit, and spousal locators provide them ways to find each other in less time and possibly with less anger. These technologies and practices form part of a complex mobile assemblage of technologies, discourses and practices for us all—for after all, we are all aging.

NOTES

1. See the most recent Canadian Wireless Telecommunication Association's study, 2011 Cell Phone Consumer Attitudes Study.
2. There are exceptions to this observation with the recent release of "Older People, Technology and Community: The Potential of Technology to Help Older People Renew or Develop Social Contacts and to Actively Engage in Their Communities," Independent Age Supporting Older People at Home, Calouste Gulbenkian Foundation, 2011.
3. The names of the participants have been changed for confidentiality purposes.

SOURCES

Canadian Wireless Telecommunications Association. "2011 Cell Phone Consumer Attitudes Study." April 29, 2011. Accessed September 1, 2011. www.cwta.ca/CWTASite/english/facts_figures_downloads/Consumer2011. pdf

Cresswell, T. *On the Move: Mobility in the Modern Western World*. New York: Routledge, 2006.

Crow, B., and K. Sawchuk. "Redressing Silences, Confronting Mobility: Seniors, Cell Phones and Aging." Social Science and Humanities Research Council. Research Grant, 263374, 2009–2011.

de Beauvoir, S. *All Said and Done*. New York: Penguin, 1974.

Gallop, J. *Anecdotal Theory*. Durham: Duke University Press, 2002.

Gitelman, L. *Always Already New: Media, History, and the Data of Culture*. Cambridge, MA: MIT Press, 2008.

Ito, M., D. Okabe and M. Matsuda. *Personal, Portable, Pedestrian: Mobile Phones in Japanese Life*. Cambridge, MA: MIT Press, 2005.

Katz, J. *Handbook of Mobile Communication Studies*. Cambridge, MA: MIT Press, 2008.

Katz, J., and M. Aakhus. *Perpetual Contact: Mobile Communication, Private Talk, Public Performance*. Cambridge, UK: Cambridge University Press, 2002.

Latour, B. *Reassembling the Social: An Introduction to Actor-Network Theory*. New York: Oxford University Press, 2007.

Lee, S. "Older Adults' User Experiences with Mobile Phones: User Cluster Identification." *Proceedings of the 21st International Symposium: Human Factors in Telecommunication, User Experience of ICTs*. Kuala Lumpur, Malaysia: 39–47, 2008.

Ling, R. *The Mobile Connection: The Cell Phone's Impact on Society*. San Francisco: Morgan Kaufman, 2004.

Ling, R. *New Tech, New Ties: How Mobile Communication Is Reshaping Social Cohesion*. Cambridge, MA: MIT Press, 2008.

Massey, D. *For Space*. London: Sage, 2005.

Mitchell, C. "The Grey Tsunami." *The Mark*. June 7, 2010. Accessed December 28, 2010. http://pioneers.themarknews.com/articles/1653-the-grey-tsunami/

Morris, M. *The Pirate's Fiancée*. London: Verso, 1988.

Morris, M. *Identity Anecdotes: Translation and Media Culture*. London: Sage, 2006.

Probyn, E. *Sexing the Self: Gendered Positions in Cultural Studies*. London: Routledge, 2004.

Sawchuk, K. "From Gloom to Boom: Age, Identity and Target Marketing." In *Images of Age*, Featherstone and Wernick (eds.). New York: Routledge, 1995, 173–187.

Sawchuk, K., and B. Crow. "Talking 'Costs': Seniors, Cell Phones and the Personal and Political Economies of Telecommunications in Canada." *Telecommunications Journal of Australia*. no. 4 (2010): 55.1–55.11.

Statistics Canada. "Aging Population." 2010. Accessed September 1, 2011. www4.hrsdc.gc.ca/.3ndic.1t.4r@-eng.jsp?iid=33

Whitney, L. "Cell Phone Subscriptions to Hit 5 Billion Globally." CNET. February 16, 2010. Accessed September 1, 2011. www.cnet.com/news/cell-phone-subscriptions-to-hit-5-billion-globally/

World Health Organization. "Active Ageing: The Concept and Rationale. Active Ageing: A Policy Framework." 2002. Accessed August 16, 2010. www.who.int/ageing/publications/active/en/

11 "I'm Melvin, a 4G Hot Spot"

Thom Swiss

The global population includes a diasporic multitude of refugees and migrants as well as a multitude of homeless individuals, families and children who lack full assimilation into their countries. "Home" for all of us is an inevitably problematic space, but to be without a home in a home-centered culture is both a traumatic experience and a defining legal state. In most countries a home address is expected—and in many cases required—for full citizenship.

Zygmunt Bauman has famously described these mobile populations as "human waste": "redundant" humans who are "superfluous" products or "side effects" of contemporary societies' quest for order and economic progress (Bauman 2007). Despite their numbers and the attention advocates have brought to these populations, the homeless, like the aging (see Crow and Sawchuk's chapter in this volume), have been left out of not only most statistical studies generated by the Internet and wireless industries, but also by most critical studies on media uses and practices.

In their January 2012 annual point-in-time ("snapshot") count, The U.S. Department of Housing and Urban Development found that 634,000 people across America were homeless. Because of turnover in the homeless population, the total number of people who experience homelessness for at least a few nights during the course of a year is known to be considerably higher than snapshot counts. A 2010 study estimated the actual number of such people to be between 2.3 million and 3.5 million. Thirty-nine percent of the homeless population is young people under eighteen, and each year over 800,000 children and youth in the United States experience homelessness (National Alliance to End Homelessness 2009).

This chapter takes up many of the themes of our book: the politics of mobility and immobility; Internet connectivity as a material practice and complex social marker; neoliberalism as governmentality; and the interplay of competing technological and social imaginaries. The chapter engages these themes by employing the logic and methods of assemblage. Its subject is homelessness with the beginning of an emphasis—assemblages can be built on and grow to any length—on homeless school-age children and youth in the United States.

I start with a large-scale event that took place in Austin, Texas, in March 2012, an "experiment" that embodies—as does Darin Barney's example in this volume's first chapter—complications around the moral valorization of mobility and the wireless Internet as well as the cultural designation of mobile access to communication networks as a expression of and near synonym for both "privilege" and "freedom." Social imaginaries, as the contributors to this volume have emphasized, are not only a set of ideas about the social world, but they are also pragmatic templates for social practice. In this way, social imaginaries, as the case below illustrates, enable particular activities and discourses and constrain others. Imaginaries provide a map of the social as a moral space.

At South by Southwest (SXSW)—an annual music and interactive media showcase that attracts fifty thousand attendees—BBH, a global advertising agency, staged what the company called "Homeless Hotspots: A Charitable Innovation." The agency's website states that BBH explores "emerging platforms and behaviors on behalf of brands, " and here is how this particular "innovation" was described by BBH:

> This year in Austin, as you wonder between locations murmuring to your coworker about how your connection sucks and you can't download/stream/tweet/instagram/ check-in, you'll notice strategically positioned individuals wearing "Homeless Hotspot" t-shirts. These are homeless individuals in the Case Management program at Front Steps Shelter. They're carrying MiFi devices. Introduce yourself, then log on to their 4G network via your phone or tablet for a quick high-quality connection. You pay what you want and whatever you give goes directly to the person that just sold you access.
>
> (BBH Global n.d.)

Not surprisingly, "Homeless Hotspots" had its defenders who, especially in brief comments on Twitter and Facebook, found the experiment "a great model" and "inspirational." Some longer-form critics also found things—more subtle things—to like. P. J. Rey writes:

> The Internet backbone—especially urban wireless infrastructure—generally exists as a series of nodes not remarked on, or massive nondescript buildings housing server farms just outside the attention of urban knowledge workers like myself. I don't need to know how it happens. The infrastructural activity that undergirds so much of my work and life goes on whether I notice it or not. What's interesting about BBH's efforts is that they bring the *infrastructure* directly into focus with mobile hotspots that you must see, name, and approach. I think the short-term publicity stunt may address the invisibility of the homeless. BBH is asking us to accept homelessness as a feature of a wireless urban landscape to be navigated.
>
> (Rey n.d.)

Newspapers, including the *New York Times*, reported on the event and the controversy surrounding the event only briefly and in a supposedly "objective" way (and in their back pages). But there was a strong backlash among some critics who considered how the experiment fit into broader political-economic currents. At the core, many comments were critiques of the strand of neoliberal ideology that prizes maximum private sector innovation. But a few also addressed the human infrastructure issues raised above. Dan Greene wrote:

> Homelessness, a clear sign of the US urbanism's structural failures, does not disappear, but is refigured into a cleaner, more productive interface for tech entrepreneurs—the new face of the city. Cross-class contact is reintroduced, but technologized and thus neutralized, shorn of risk or surprise . . . It's not merely that homeless people fail to directly benefit from the thing they are being put to work producing but that this form of wage labor, guised as charity, also happens to be completely unregulated. That is to say, the homeless hotspots scheme takes exploitation to extremes. What are the consequences of living a society that acquiesces the fact that its constituents are not even guaranteed $60 for a hard day's work because their service is classified as charity and not as proper labor? What happens to the already vastly disproportionate distribution of wealth if companies can supplant their workforce with so-called charity cases?
>
> (Greene n.d.)

Debates central to *Theories of the Mobile Internet: Materialities and Imaginaries* are present in these and other comments about "Homeless Hotspots: A Charitable Innovation." As part of a new mobilities paradigm, these debates over the mobile Internet continue to generate new discussions and provisional answers to questions that are themselves new.

I want to take up here one of the conceptual touchstones for *Theories of the Mobile Internet*: the conceit of assemblages. I'll start by noting that most of this chapter *is* an assemblage or the start at an assemblage that experiments with the form of the academic article. It does not proceed in traditionally linear fashion and does not make an explicit argument. Instead, it engages the contingent, the associative and the intertextual by linking items that include newspaper reports, policy statements, popular representations, academic writing, lyrics to a song and other materials I've collected in the last few years. The chapter does so in part because I believe a piecemeal, fragmented approach to writing mirrors my topic: the fragmented lives of the homeless. But there are more important reasons as well.

Among the tutor texts for my approach to writing this chapter is John Law's (2004) book, *After Method: Mess in Social Science Research*, in which he argues that assemblages are ways of "apprehending or appreciating displacement. Each fragment is a possible image of the world, of our experience of the world, but so too is their combination." Manuel DeLanda's work, building on Deleuze's theory of assemblages, also informs my approach to writing this chapter. DeLanda argues that the components in assemblages are defined by their material dimension and territorializing and deterritorializing axis. "The components in assemblages," he has noted," are historically contingent, heterogeneous and self-subsistent, giving the possibility to take one assemblage and insert it into another without destroying its identity. The main characteristics of the relationship between an assemblage and its components are complexity and non-linearity" (DeLanda 2006). Finally, two additional sources offered *stylistic* models: Roland Barthes's late work—in which he deploys fragments as a series of interruptions with cumulative intellectual and aesthetic surprises—and literary and visual assemblage*s* as both artistic processes and products in which new images and texts are primarily and explicitly constructed from existing ones.

It is hardly news to say that over the last few decades interdisciplinary methods previously thought of as peripheral to "serious" academic research have become increasingly visible in many fields. Language- and image-based stylistic interventions appear occasionally in our academic journals, including poems, personal narratives and photographs intended to function as arguments or critiques. Following Barthes (1975) then, who argued that one cannot produce new ways of doing criticism without changing the structures that govern it, and Law, who argues "we will need to teach ourselves to know some of the realities of the world using methods usual to or unknown in social science," my writing here *begins* an assemblage that could grow, if space allowed, in a number of different directions. Like many assemblages of the kind I am describing, this one on homelessness and digital connectivity is provisional and generative; it does not intend to make a specific argument but rather to serve as an alternative heuristic.

My *ad hoc* process of collecting, selecting, arranging, titling and editing multiple representations of and discourses about "homelessness" and "mobility" over time is meant to emphasize the *ad hoc* and active meanings of these terms and to help articulate the range of actors, concepts, practices, and relations at work in the production of these terms as concepts. The texts I reproduce and remix demonstrate that the meanings of individual fragments are constructed not only within the social and media-specific context in which they first appeared, but also through their relationships with other texts that resonate, sometimes broadly, within culture. Assemblages are emergent; in what follows, individual fragments are overlapping and iterative while the structure of the whole attempts to pay attention to unintended effects.

I've generally titled the sections of my assemblage by borrowing a word or phrase from the passage that follows, except when the passage seemed to deny me a summative or provocative title. In those cases, I've invented my own. Many of the fragments are verbatim, although typically snipped from larger blocks of prose, and some are edited for clarification by the internal logic of this assemblage which is, as a whole, guided by a spirit of *poesis*.

Wireless mobile Internet connectivity brings profound changes to many aspects of life, including in some cases greater inequality as well as more social and economic polarization. Homeless individuals are sites traversed by complicated and often competing discursive paths; they are also human subjects, bodies routed and rerouted, directed and misdirected by circumstances every day. How does digital connectivity and the wireless Internet change these circumstances? In a few of the following passages, themes related to a specific place (Minneapolis, Minnesota) begin to emerge in this assemblage. Minneapolis figures for several reasons: I live in the city's downtown area and urban homelessness is visible to me. I read frequent accounts in the local media about both homelessness and the city's wireless infrastructure. Finally, the Minneapolis public school district has been recognized by *Time* magazine and other sources as a national "model" for school districts struggling with meeting the needs, including the wireless Internet needs, of homeless students. Ironically, in the context of this chapter, these young people are often officially referred to as "highly mobile" students in policy and legal documents.

"I'M MELVIN, A 4G HOT SPOT": AN ASSEMBLAGE

"Within the Surround of Force"

Poverty means being excluded from whatever passes for normal life. This results in loss of self-esteem. Poverty also means being cut off from the chances of whatever passes in a given society for a happy life. This results in resentment.

In a consumer society a normal life is the life of consumers, preoccupied with making their choices among the panoply of publicly displayed opportunities for pleasurable sensations and lively experiences. A happy life is defined by catching many opportunities, catching the opportunities most talked about and thus most desired.

Those in poverty are unable to participate as fully fledged consumers. They are marginalized and made to feel inadequate. They live in a surround of force that differentiates them from those inside the circle of power.

Within the surround of force, people live in panic. They scurry, going from place to place, looking for food, a new apartment, medical care for a child. The wall of the surround pens them into a limited area, but the panic inside the surround has no limits; people may do nothing and everything, suffering from excesses of both order and liberty.

In other words there are no constants within the surround, no reason, no stability, and no rules; there is only force (Bauman 2007).

"Fixed in Mobility"

When researching the relationship between mobility and homelessness it is important to consider how fixity is not always mobility's opposite, rather some people become *fixed in mobility* (Jackson 2012).

"Two Million More"

The National Alliance to End Homelessness estimates that, because of the recession, as many as a two million additional Americans have become homeless in the last two years. For lower-income working families, it means one poor decision can rapidly deteriorate into a maelstrom of debt and financial problems.

Many of the newly homeless do not fit *the stereotype of homelessness*. They may be hard-working, healthy and addiction-free (National Alliance to End Homelessness 2009).

"Tweeted at 6:41 PM—8 Mar 2012, Austin TX"

Homeless Hotspots@HHSXSW
Our Hotspot Managers are excited (and a bit nervous), so if you're in town tomorrow come say hi and make them feel at home. #SxSW
(BBH Global n.d.)

"Survival Devices"

For the homeless, mobile devices are not viewed as being a tool for recreation or a status symbol, but *a survival device*. The media talks about teens, young adults, moms and even senior citizens engaging in social media, but little is ever discussed around the homeless and how they are using digital technology such as mobile or social media.

People turn to Facebook to post their gripes; they receive support and advice from friends and others who can relate. For the homeless, social sites such as Facebook, Twitter and even email are used to maintain life-saving connections. For the homeless, a simple cell phone can be a vital link to a family member, support group, employment and housing opportunities, and is a way to build a community out of the isolation, ruin and despair that accompanies homelessness. A mobile phone offers the disenfranchised a cheap way to communicate, basic Internet access can connect them to a wealth of information and resources (Johns n.d.).

"Giant Welfare State"

During the March 6 broadcast of his nationally syndicated radio show, Rush Limbaugh, discussing a photo of Michelle Obama's March 5 visit to a Washington, DC, homeless shelter that featured a man photographing

Obama using a cell phone, said that "the homeless and the poor are show-ing up taking pictures of her with their cell phones." Limbaugh later added:

> What we have, ladies and gentlemen, is a giant welfare state that's in the process of being manufactured. Like, you're going to have to see the first lady behind the counter at McDonald's when you go in there as your poverty-stricken day drags on—take a picture with your cell phone while you go in there and get your McNuggets or whatever's being handed out that day.
>
> (Bruck n.d.)

"People Who Care about Me Are on Facebook"

A USC School of Social Work study revealed three-in-five homeless teens have cell phones and wireless access.

> "Why can't I be on Facebook?" asked one subject in the study. "I have as much right to that as anyone else. Just because I am homeless does not mean that I don't care about this stuff, you know? My family is on Facebook. My friends are on Facebook. People who care about me are on Facebook."
>
> (Johns n.d.)

"Obama Phones"

A federal program designed to help homeless and other impoverished peo-ple connect with family, friends, housing programs and potential employers could provide potentially millions of poor people with free cell phones and service.

The program has come under fire from some Republicans, who cite it as an example of government largesse. Opponents have labeled the devices "Obama phones," even though the initial program was created under Presi-dent Ronald Reagan and expanded to include cell phones during the George W. Bush administration. Congressman Tim Griffin (R-Ark.) recently intro-duced a bill aimed at gutting the program (Hubert n.d.).

"Mobilities Paradigm"

Whereas the new mobilities paradigm offers a rather loose framework for scholars seeking to engage with mobility research, there have been recent attempts at greater theorization. Tim Cresswell's work possibly represents the most sophisticated theoretical endeavor.

Cresswell posits that mobility has three interconnected dimensions—movement, meaning and practice—which combine into different "constel-lations of mobility" and shape the "politics of mobility."

First of all, mobilities are about movement. Movement is closely related to place, as mobility happens in places and through places. Movement is itself made of different dimensions that Cresswell identifies as purpose, velocity, rhythm, route and spatial scale.

Second, mobilities are meaningful, that is, they do not take place in a vacuum but in socially and culturally constructed systems of meaning. Mobilities mean different things to different people in different societal, cultural and historical contexts. For instance, the same journey between two specific locations acquires very different meanings in different contexts such as tourism and immigration. In other words, mobilities are a relational phenomenon.

Third, mobilities are practiced. This means that the experience of movement may be extremely different depending on a number of factors. Under diverse circumstances, *the practice of moving may range from being an exhilarating experience to being a boring routine or a life-threatening adventure*. Mobilities are an experiential phenomenon (Söderström 2010).

"Oh, God, That's So Inspiring—You Got Your Clothes from a Garbage Can"

A twenty-eight-page pictorial in the September issue of *W* magazine, shot by the British photographer Craig McDean, repurposes shopping bags from labels like Chanel and Dior as makeshift dresses, and shows them worn with furs and pearls and designer bags. As she prepared for her debut runway show on Friday, Erin Wasson, a model turned designer, seconded Mr. Duffy's view that the professionals could take some tips from the homeless, and she defended remarks she made last year in an interview with *Nylon* magazine: "The people with the best style for me are the people that are the poorest." "It's not like I'm saying, 'Oh, God, that's so inspiring—you got your clothes from a garbage can,'" Ms. Wasson said.

What is she saying then? "When I moved down to Venice Beach, I found these people with this amazing mentality, this gypsy mentality—people that you couldn't label and put in a box" (Trebay 2009).

"Movement Belongs to the Privileged Whereas the Poor Are Imagined as Living Slower"

Drawing on Raymond Williams's terminology, we note the emergence of a mobilities "structure of feeling." But those who have emphasized global flows have been criticized for neglecting the "short haul." The importance of everyday, slower movement is inescapable when considering the lives of young homeless people. Massey argues: "Much of life, for many people, even in the heart of the First World, still consists of waiting with your shopping for a bus that never comes."

But this critique risks reinforcing the idea that movement belongs to the privileged whereas the poor are imagined as living slower, more fixed lives. Within this schema, distance from global flows equals disadvantage, *and yet, strangely, homelessness is often also associated with perpetual movement* (American Sociological Association 2012).

"The Living Homeless"

In "Night of the Living Homeless," an episode of Comedy Central's *South Park*, homeless people have been showing up in South Park in large numbers. The town council has taken notice of the problem and come up with some solutions—turning the homeless into tires for their cars or giving them designer sleeping bags and makeovers "so they would at least be pleasant to look at" (Wikipedia 2014).

"A Yelp-Like Site"

A computer lab inside a homeless services nonprofit is not the usual place for a tech industry press conference, but it is sobering, and it may be the most powerful. Downstairs, the rain poured down intermittently outside as homeless and low-income people waited in line to eat.

San Francisco's officials want the mobile site to be a symbol of what could be in a place where tech, and those who have supported tech businesses through tax cuts, is vilified.

The Link-SF site, accessible on the Web but optimized for mobile phones, is pegged as a "Yelp-like" site for homeless services. Users can look up resources for food, shelter, medical services and places to access showers or technology (Tam 2014).

"Who Should Be Counted as Homeless"

With unemployment and foreclosures rising and growing numbers of families struggling to find affordable housing, lawmakers in Congress are debating who should be counted as homeless. For more than twenty years, federal housing law has counted as homeless only people living on the streets or in shelters.

In the House, which is expected to vote on the issue in May, lawmakers are discussing whether to expand the definition to include children and their families in desperate need of stable housing—about one million people—or to add a much smaller group that would include only people fleeing their homes because of domestic violence and those who can prove they will lose their housing within fourteen days.

The Senate is considering a still narrower expansion that would include only those who have been forced to move three times in one year or twice in twenty-one days. None of the bills come with any additional financing (Swarns 2008).

"Homeless"

> Zio yami, zio yami, nhliziyo yami
> Nhliziyo yami amakhaza asengi bulele
> Homeless, homeless
>   Moonlight sleeping on a midnight lake
>
> (Simon 1986)

During the last few decades, the face of homelessness has changed as families with children have begun to use emergency shelters in the Twin Cities and across the country. School access is known to be a major problem for highly mobile students and children lacking a stable home address due to residency requirements, lack of transportation and missing records (Kingsbury 2009).

"Homeless for 10 Months. Courted by Online For-Profit College."

Benson Rollins wants a college degree. The unemployed high school dropout who attends Alcoholics Anonymous and has been homeless for ten months is being courted by the University of Phoenix. Two of its recruiters got themselves invited to a Cleveland shelter last October and pitched the advantages of going to the country's largest for-profit college to seventy destitute men.

Rollins's experience is increasingly common. The boom in for-profit online education, driven by a political consensus that all Americans need more than a high school diploma, has intensified efforts to recruit the homeless. Such disadvantaged students are desirable because they qualify for federal grants and loans, which are largely responsible for the prosperity of for-profit colleges (Golden 2010).

"Minneapolis"

Right now, nearly one in ten children attending public school in Minneapolis is homeless. Read that sentence again.

As Wall Street tries to right itself, the global economic crisis is punishing many of the youngest Americans. Preliminary nationwide figures indicate that there were nearly 16 percent more homeless students in the 2007–08 academic year than in the previous year, and the number of homeless students continues to climb as more parents face foreclosure or the unemployment line. Over the past two decades, Minneapolis's thirty-three thousand-student district has seen a steady increase in the number of homeless kids, as the

Twin Cities area has hemorrhaged manufacturing jobs and the supply of affordable housing has dwindled (Kingsbury 2009).

"Assemblage Theory"

Assemblage theory is an approach to systems analysis that emphasizes fluidity, exchangeability and multiple functionalities. Assemblages appear to be functioning as a whole, but are actually coherent bits of a system whose components can be "yanked" out of one system, "plugged" into another and still work. As such, assemblages characteristically have functional capacities but do not have a function—that is, they are not designed to only do one thing (Texas Theory Wiki n.d.).

SOURCES

American Sociological Association. "Study: Homeless people find equality, acceptance on social networking sites." 2012. Accessed June 19, 2013. http://phys.org/news/2012–08-homeless-people-equality-social-networking.html#jCp

Barthes, Roland. *S/Z: An Essay*. London: Hill and Wang, 1975.

Bauman, Z. *Work, Consumerism and the New Poor*. Buckingham, UK: Open University Press, 1998.

Bauman, Z. *Liquid Times: Living in an Age of Uncertainty*. London: Polity, 2007.

BBH Global. Accessed July 20, 2014. http://www.bartleboglehegarty.com/newyork/

Bruck, L. Accessed August 15, 2014. http://mediamatters.org/research/2009/03/06/limbaugh-runs-with-homeless-cell-phone-smear/148062

DeLanda, M. *A New Philosophy of Society: Assemblage Theory and Social Complexity*. London: Bloomsbury, 2006.

Golden, D. "The Homeless at College." *Bloomberg Business Week*. April 30, 2010. Accessed October 20, 2011. www.businessweek.com/magazine/content/10_19/b4177064219731.htm?chan = magazine+channel_features

Greene, D. "Urban Mobility: Homeless Hotspots and ICT4D." *Cyborgology*. Accessed March 10, 2014. http://thesocietypages.org/cyborgology/?s = homeless+hotspots

Hubert, C. "Free Cellphones For Homeless." *Huffington Post*. June 19, 2013. www.huffingtonpost.com/2012/12/14/free-cellphones-for-homeless_n_2298895.html

Jackson, E. "Fixed in Mobility: Young Homeless People and the City." *International Journal of Urban and Regional Research*. no. 22 (2012), 12–35.

Johns, M. "US Homeless Turn to Smartphones and iPads to Survive." Accessed February 1, 2014. www.business2community.com/tech-gadgets/the-us-homeless-turn-to-smartphones-and-ipads-to-survive-0375464#oqGaSTcqMtyadqkv.99

Kingsbury, K. "Keeping Homeless Kids in School." *Time*. March 12, 2009. Accessed March 28, 2010. www.time.com/time/magazine/article/0,9171,1884822,00.html

Law, J. *After Method: Mess in Social Science Research*. London: Routledge, 2004.

National Alliance to End Homelessness. *2009 Annual Report*. Accessed February 1, 2014. www.endhomelessness.org/library/entry/increases-in-homelessness-on-the-horizon

Ode, K. "American Girl Unveils Homeless Doll." *Baltimore Sun*. 1998. Accessed December 2, 2011. www.baltimoresun.com/features/parenting/bal-american-girl-gwen-0930,0,5319017.story

Rey, P. J. "Homeless Hotspots: Branding Masquerading as Charity." *Cyborgology*. Accessed March 10, 2014. http://thesocietypages.org/cyborgology/2012/03/13/homeless-hotspots-exploitation-masquerading-as-charity-at-sxsw/

Shorris, E. 2000. *Riches for the Poor—The Clemente Course in the Humanities.* London: W.W. Norton and Co, 2000.

Simon, Paul. *Graceland*. New York: Warner Brothers Records, 1986.

Söderström, O. "The Constitution of Society: Rethinking the Mobility/Society Nexus." Working paper, 2010. Accessed March 1, 2014. https://doc.rero.ch/record/19815/files/S_derstr_m_Ola_-_The_mobile_constitution_of_society_rethinking_the_mobility_20100630.pdf

Swarns, R. "Capitol Strives to Define Homeless." *The New York Time*s. September 16. 2008. Accessed January 2, 2014. www.nytimes.com/2008/09/16/washington/16homeless.html?pagewanted = all&_r = 0

Tam, D. "Tech Doing Good? SF Tackles Homelessness with Smartphones," *C/net*. February 28, 2014. Accessed March 14, 2014. http://news.cnet.com/8301–1035_3–57619741–94/tech-doing-good-sf-tackles-homelessness-with-smartphones/

Texas Theory Wiki. "Assemblage Theory." Accessed March 2, 2014. http://wikis.la.utexas.edu/theory/page/assemblage-theory

Trebay, G. "Aware of the Homeless? Well, You Could Say That." *The New York Times*. September 12, 2009. Accessed June 1, 2010. www.nytimes.com/2009/09/12/fashion/12DIARY.html

Wikipedia. "Night of the Living Homeless." Accessed March 13, 2014. http://en.wikipedia.org/wiki/Night_of_the_Living_Homeless

12 A Hole in the Hand
Assemblages of Attention and Mobile Screens

J. Macgregor Wise

Technology, precisely in its miniaturization—the whole digital world in your hand—acquires a magical quality.

Michael Bull

We are closer here in effect to the tactile than to the visual universe.

Jean Baudrillard

THE SWARMING OF THE ASSEMBLAGES

This chapter begins with a swarm of screens.[1] Or I should say, the swarming of screens. I have in mind the intersection of Michel Foucault's (1977) notion of the swarming of disciplinary mechanisms, Howard Rheingold's (2003) characterization in *Smart Mobs* of the swarming of the new networks of communication and computation and Michel de Certeau's (1998) observation of the "swarming structures of the street." These turns of phrase also echo the language of Deleuze and Guattari and it is they whom I will follow in the ensuing pages. In these swarms, that is, I hear the murmurs of new assemblages (1986).

The swarming screens I have in mind are, on the one hand, fixed and immobilized into our ambient architecture, as Anna McCarthy (2001) pointed out a decade ago. As we become mobile, screens come to where we are: restaurants, malls, elevators, gas pumps, grocery stores, automobiles, waiting rooms of all sorts, and so on. But, on the other hand, there are screens (often referred to as the Third Screen, Nicholson 2010) which themselves are mobile, carried in the pocket, part of our mobile phones, PDAs, or similar devices, and now embedded in glasses. They are increasingly ubiquitous and banal.

The ways that these swarms have been understood have been primarily in terms of connection and distraction (cf. Rheingold 2012). They are part of our practice of constant contact with absent others. The exchange of mobile phone images, for example, is one way people stay in synch with friends and significant others, a way of confirming a moment of sociality.

They also share in the processes of association or socializing characteristic of other mass media like radio and television (when watching, we are being social [cf. Winocur 2005]), and they are seen as yet another factor eroding our attention, slicing it into ever thinner moments as we leap, dizzy and distracted, from text to to webpage to Angry Birds (Gallagher 2009, Jackson 2008). Cognitive scientists worry that filling up every second with something to do (now that we have our portable distraction factory in our hands) erodes our ability to form memories or to learn (Richtel 2010).

In this chapter I approach the swarm of mobile screens from a different angle. What if we see them as *assemblages*, following Deleuze and Guattari's use of the term (1987, see also Wise 2011), and locate those assemblages within a broader shift in assemblages, to something I have been calling the Clickable World (Wise 2012a). Put simply, the Clickable World is a social imaginary that posits that the ways that one navigates online social spaces is becoming how one navigates one's "real" life. The world appears malleable, available, interactive and information-filled. The Clickable World is where one approaches the everyday environment as being analogous to navigating the Web on a computer screen: just click around for further information. There is a presumption of agency here in this imaginary, human agency over the environment—the world is at hand, in control.

The Clickable World assemblage emerges as we transition from the disciplinary assemblage mapped by Foucault to the assemblages of control suggested by Deleuze in his later writings (1995). In a disciplinary society, power shifted from being corporeal, focused on physical domination of the body, to incorporeal, power through knowledge. Famously in the Panopticon, discipline is engrained through vision and knowledge of the prisoner. Foucault then described the swarming of disciplinary mechanisms beyond the confines of their institutional sites (to surveil workers or patients or students in their homes as well as at work, the hospital, and so on). The idea of a society of control sees the further dissolution of institutions to a constant continuous application of control. But unlike a disciplinary regime's reliance on "optical visuality," Inke Arns (2013), drawing on Laura Marks (2004), has argued that "post optical regimes" rely on "haptic visuality." Marks grounds her approach in Deleuze and Guattari's notion of smooth space, which is "a space that must be moved through by constant reference to the immediate environment, as when navigating an expanse of snow or sand" (Marks 2002, xii; 2004, 80). Control is constant and close-by, worn in devices, embedded in the environment; it is corporeal, material and mobile. In mapping the materialities and imaginaries of the mobile Internet, I would ask that we pay attention to attention—to attention as corporeal and haptic—to trace the articulations of swarming attentional processes, orientations, habits and devices, to bring them back to visibility. Attention reveals important aspects of agency and control if care is taken not to valorize either the agency of the attentive human agent or their lack of agency in the face of new swarms of devices.

In particular I want to consider mobile screens within an assemblage of attention characteristic of the Clickable World. The assemblage idea of attention is about neither the proliferation and consequences of distraction on everyday life (cf. Jackson 2008) nor the emergence of economies of attention (cf. Beller 2006; Lanham 2006, but see also Terranova 2012). Rather, it is the distribution of attention, or attentional processes, across brain, body and environment (Wise 2012a). Attention in the new assemblage, as I argue elsewhere,

> cannot just be about the particular configurations of our individual cognition (e.g., deep or hyper attention), but about attentional processes scattered across our devices. Perhaps this is what's new here—this contemporary assemblage (beyond brain, skin and human), more than others, is one that *attends* (2012a, 168).

> A focus on assemblages of attention means that we need to draw the line, make the connections, between devices which pay attention to us (the surveillant assemblage), devices which seek to manage our attention (to both attract/distract us and also track our attention so as to better attract/distract) and our own cognitive and habitual attentional processes. That is, to speak of attention today is to speak of the contingent aggregation, articulation and expression of all of these processes (2012a, 169).

In this chapter I explore one instance of this attentional assemblage, that of mobile screens. Obviously mobile screens are particularly rich and intense points of gravitational pull on our attention. Everywhere we turn, it seems, someone is attending to a screen.[2] However, we should also note, given the above discussion, that these devices themselves attend—to their position in space (if it has an accelerometer), to Wi-Fi or cell phone signals, to the proximity of other devices.

Mobile screens themselves provide a generative case with which to explore the implications of this assemblage, but I will focus on an even more specific case, and that is the viewing of live webcam images on mobile devices through applications like LiveCams. My chapter continues work I did a decade ago on webcams and everyday life (Wise 2004). It also moves away from just focusing on the screen in hand to include the networks through which it is articulated and the remote locations depicted.

NEW SCREENS

Donna Haraway (2008) has a wonderful essay on *Crittercam*, a National Geographic show where cameras are attached to animals in the wild. Following the work of Don Ihde, she talks about the infoldings of technology and animal, and also viewer and screen, and of course the asymmetrical infoldings of viewer, technology and animal. The implication of this assemblage (my term, not hers) is, for Haraway, "epistemological-ethical obligations to the animals" (263). I don't want to rehearse Haraway's subtle analysis here, but point to three aspects of mobile screens implied by her analysis. First, we need to consider the relation of self to screen, in this case, the screen in the hand, both furtive and controlling (us watching). Secondly, we need to consider the materiality of the arrangement of camera and subject. The camera is embedded in a landscape; the camera is surveilling a place of business or leisure, with broader implications and articulations to state-sponsored civic traffic infrastructure, organizational self-surveillance, and so on. Third, the articulations through networks of communication and computing that link one to the other imply an ethical responsibility. This means that in viewing these images I am participating in the surveillance at the other end. These three aspects are a part of this assemblage of attention: our attention to the screen in my hand, the attentions of the remote camera, and the myriad devices, processes and relationships established to articulate one to the other.

Body-to-Screen

A Screen in the Hand

We begin with the infoldings of hand and screen, and the production of the experience of attending to the screen. What has been called the phenomenology of the mobile screen is an expression of this assemblage.

In 2000, while walking in Tokyo, writer Howard Rheingold (2003) recognized a new shift in the technological everyday when he realized that some pedestrians weren't talking on their mobile phones, but staring at them. In my own terms, he was noting a shift in the technological assemblage, especially urban street culture. The urban pedestrian has a particular way of being in and negotiating the space of the city street that is always technologically inflected. Such assemblages are also continually shifting with social, cultural, economic and technological transformations. From the flâneur of the nineteenth century assaying the newly electrically illuminated street, to the privatization of public space made possible by the Walkman in the 1980s, to the opening up of local spaces to elsewheres through the constant communication of the mobile phone, each articulates bodies, technologies, epistemologies and phenomenologies: ways of embodying space. The altered relation of the subjects with their environmental contexts can be noted by changes in gesture, position and attitude. Pedestrians with portable music players stare into space, not making contact with those around them, an attitude continued by people on their mobile phones (attending to a space neither here nor there, but the space of the phone call) (cf. Bull 2007). The attentional gravity of a phone call is to the space of the call which is always other than the physical space that we are in, though we do vaguely attend to that as well. It is not a question of either/or (mediated space or physical space) but the grasping of both, like a form of stereo vision, what Paul Virilio (2000) called *stereo-reality.*[3] What Rheingold noticed was rather than people walking with one hand clasping a phone to their ear, people were *looking* at their phones. The phone itself becomes a strong point of visual attentional gravity. These people were, of course, texting: typing and reading short messages. Rheingold was noticing this new behavior in its early days, before it had become habit. Soon texting was to become habitual and haptic, that is, performed through feel, not vision. Though the phone in the hand remains a crucial component of the contemporary pedestrian (indeed, they tend to be constantly in our hands like large worry beads—Plant [2001]), they eventually became more the subject of our haptic attention than our visual attention. However, as Cooley has argued, haptic and visual attention are not mutually exclusive, indeed the phone in the hand seems to transform what and how we see (Cooley 2004).

The questions of this chapter are as follows: if one function of a television screen is to look through it (rather than at it); that is, if a live television image transports us phenomenologically to an elsewhere (in a way quite different from the aural elsewhere of a phone conversation), what do we make of the phenomenon of live video images held in our palms, the result of mobile television and webcam applications? There is a hole in one's hand, and how does the hole in one's hand implicate us in extended assemblages of care and control?

The phenomenology of mobile media devices is being mapped by Ingrid Richardson (Richardson 2005; 2007; 2011; 2012; Richardson and Wilken 2009) and others (see, e.g., Farman 2012). So let me touch on some relevant findings. Note, however, that most of these scholars take mobile locative gaming as their object of study, rather than webcams or video viewing. Richardson observes that these screens are only glanced at, usually used in the context of other activities, momentary and interruptible. Content for mobile television, as a Nokia report put it, will be "snackable" (Orgad 2006), designed to fill those distracted, in-between moments. The notion of a person consuming media in distraction is not a new one, by any means, and certainly not unique to these devices—though, perhaps, there are a greater number of sources of distraction today. Walter Benjamin and Sigfried Kracauer noted in the early decades of the twentieth century that reception in distraction was one of the hallmarks of modernity (Highmore 2010). Distracted consumption (which means both a fragmented distracted object of consumption and the scattered attentional state of the receiver) is said to short-circuit critique and thought—we're not concentrating, after all (cf. Benjamin). But also, it is said to provide opportunities for critique. Reception in distraction tends to sit at the level of habit, the everyday, more a part of the rhythms of our bodies, their spaces, and their days than consciously decoded meanings and communications (see. e.g., Bull 2007; Highmore 2005; Lefebvre 2004).

Richardson also notes the intimate nature of hand-tool relation—mobile phones, for example, are habitually held; they are extensions of our hands, part of our body schema. These screens are not just visual and aural, but haptic. A key way we interface with the phone is by touch—not just dialing numbers but by texting. Larissa Hjorth argues that the "haptic has often been under-theorized in mobile communication discourses" (2009, 145).

Studies of other mobile devices have examined further this relationship of hand and screen. What Heidi Rae Cooley (2004) has termed "tactile vision," Nanna Verhoeff (2009) refers to as "haptic visuality" in her study of the Nintendo DS (dual screen) mobile gaming system. In the DS one manipulates the image and gameplay by touching a second screen with finger or stylus (comparable, though more intimate, to the mouse-screen interface of the desktop—the mouse is further removed from the action than the finger or stylus.)

A step past the DS is the direct manipulation of the image on screen by touching that screen (and not a secondary one). This, of course, is the iPod/iPhone/iPad interface. What differentiates the haptic interface of the iPhone from that of previous mobile devices is first the relation to the hand—it fits less comfortably than rounder, smaller mobile phones. More crucially, however, is the lack of buttons (beyond the one). Most interaction is done via the screen. But the screen could display anything–numbers, a piano keyboard, or playing cards. That is, unlike previous phones described by Richardson, the affordances of the iPhone *require* us to look—the surface feels the same and there is no tactile differentiation between applications (which, a contributor to this volume, Gerard Goggin [2006], has pointed out is an issue for those who cannot see).

A key feature of the device is its visually directed haptic interaction. Icons are moved, images stretched, objects manipulated by directly touching them on screen, not secondhand via a stylus, mouse, keyboard or secondary screen. We should not underestimate the sense of control this brings. There seems to be no intermediary. Even though we get used to the spatial dislocation of mouse and screen, and directly manipulate images in front of us on a desktop computer by moving the mouse at some remove from that image, it is much more powerful to reach out and (seem to) do it ourselves with our own hands. This particular sense of agency is an important part of this new assemblage.

We seem to be poised to take another step—the screen moving out of the hand and up to the eye itself, in the form of Google Glass (cf. Wise 2012b, 2013). Control of Glass is hands-free, for the most part, relying on verbal commands and head motions, and potentially even blink patterns. The device overlays a small screen in the corner of your vision. Screen and environment are both constantly present, a literal version of Virilio's stereo-reality. The phenomenology of such devices is too new to speculate (for clues, see Steve Mann's work [2001], since he's been experiencing this for decades), but the transparency of video punches a hole not in one's hand, but in the world.

Seeing Through

Bolter and Grusin (2000) differentiate between transparent media and hypermedia. *Transparent media* are those we seem to see through; the devices themselves disappear. Many of our common communication media such as telephones and televisions are transparent. *Hypermedia*, on the other hand, are displays that draw attention to themselves—we delight in the virtuosity and spectacle of the surface, but don't expect to be transported through to anywhere else. We look *at* them, not *through*. Many of the joys of the iPhone and iPod touch are precisely apps that are hypermediate. The point is to play with the surface whether it's manipulating a photograph or moving an image of a paddle in an air hockey game.

Verhoeff likewise notes the transparency of some screens, but contrasts them with opaque media. What's important for Verhoeff about opacity is that it emphasizes—following Bill Brown's work—the *thingness* of the screen, the materiality of the device. The thingness of the screen allows our relation with the device to remain haptic and habitual, but resists the device's complete disappearance into transparency.

There are a number of dangers when devices disappear. For example, we lose the ability to interrogate their place in our lives once they become taken for granted, part of the woodwork (Wise 1997). We will consider these implications toward the end of the chapter.

Lefebvre's Windows

These properties of the mobile screen (including consumption in distraction, relative privacy, in the hand/under control, visually engaging, prone to hypermediacy) depend on the practices in which these devices are involved, the types of images engaged with, and so on, rather than the device in and of itself. So, for example, webcam applications differ from, say, Angry Birds, in that they are applications of transparency, not opacity. Hence, my title: a hole in the hand. When one looks at a live, streaming webcam image, the screen in one's palm provides a view of elsewhere.

There are currently dozens of webcam applications for mobile phones. Let me give you an example of one webcam application, LiveCams.[4] The opening screen of the LiveCams app presents you with an array of twelve squares, each a miniature television screen or monitor. Each square is a randomly selected live streaming image. Several more pages of live cameras follow. One can touch an image and it fills the screen. Many give the viewer the ability to pan, tilt and zoom the distant camera. One can also search for particular places and bookmark one's favorite cams.

Unlike the more notorious of the early webcams, most of the cameras here are public cams—municipal (traffic, street scenes, beaches) or private-public spaces like skating rinks and hotel lobbies. Like early webcams, the images are blurry and indistinct, lacking detail.[5] Some of the images are eerie, almost ghostly (shadowly figures glide past in a Northern European skating rink). Generically, the shadowy images remind me of evidence from crime scenes, grainy surveillance footage shown on the nightly news. So, like early webcams, the significance of these images is *the fact that* we're seeing more than *what* we're seeing. It's about the fact that I can watch traffic in downtown Prague and not, say, the significance of particular vehicles (though this might be different if I lived in Prague and used the app to warn me about my commute). Given the quality (or lack of quality) of the image, what we see is often a product of our desire. The image is semiotically constructed by the title, place names anchoring the meaning of the image (Barthes 1977) as much as anchoring its location on the globe.

Each of the images in Figure 12.1 was in motion—the lights of cars jerkily move down a street (now here, now farther along, like flipbook images or crude animation). Some images you have to scrutinize for signs of life—or, actually, signs of liveness—in order to understand that this is, indeed, a live scene and not a screen capture. Staring at a lonely town corner, one sees a branch move slightly, or a shadowy figure step into the pool of light of a streetlamp before fading back into darkness.

Frustrated with the lack of detail, the semiotic obtuseness of the images, you tap on one and it fills the small screen. For many images, the enlargement makes a difference. For others, the blurs from ill-adjusted cameras at the limits of their resolution and bandwidth become just larger blurs. But for most, figures become more distinct, their movements fluid. The scene resolves itself into something readily identifiable. Ships move across the harbor in Japan, bikers attempt tricks in a park in Seattle, a guest confers with staff at a hotel in Russia, cars move slowly down a Swedish street. As Daniel Palmer (2000) once noted of webcams, one function is the affirmation of one's longitude in relation to the rest of the globe. Late afternoon in Phoenix, Arizona, but night in Sweden and early morning in Japan.

But then, that's it. A few more cars, another skater gives it a go, a flag moves lazily. A key characteristic of the webcam image is precisely that very little happens. As I wrote a number of years ago, webcams in general allow for a certain scrutiny of the everyday:

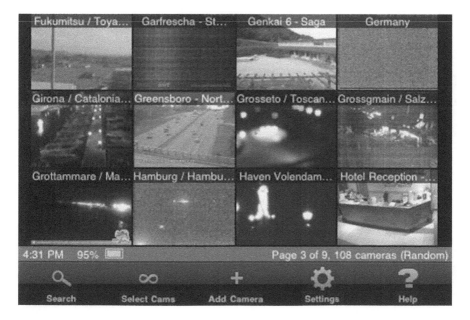

Figure 12.1   Screenshot of the app LiveCams.

[T]his scrutiny is not at the level of the close-up or enlargement (such as in film or photography), or a result of an increase in detail (because of the limitations of the medium, there is sufficient loss of detail), but a product of time. One the one hand, we have the momentary, the moment caught on camera, or a succession of such moments (a certain this-ness, this moment and then this moment) . . . On the other hand, we have the long stretches of time where we can watch a room or street or landscape for hours.

(Wise 2004, 426, 427)

The webcam becomes Lefebvre's window, a perch to study the rhythms of the everyday (2004). But this, essentially, leaves us passive, waiting for something to happen. If we so wish, we step further into the panopticon's guard tower and push the Control button (Figure 12.2). We zoom in a bit, exploring a detail. We pan to the left. More buildings? The open bay?

Unlike deskbound webcams, this mobile screen is not a medium to study the duration of everyday life (and actually makes a poor substitute for Lefebvre's window) for two reasons. First, like other mobile devices they are consumed in distraction, and so lack the ability to endure over time. They are the snack, the respite, the amusement between other amusements, activities.[6] Second, they drain battery power like nobody's business, meaning that time is very limited on the app unless you are plugged into an outlet (which defeats the purpose of a mobile device). Note the drop in battery power between Figures 12.3, 12.4 and 12.5, indicated by the battery icon in the upper right.

Figure 12.2   Japan.

Figure 12.3 USA.

Figure 12.4 Sweden.

Figure 12.5   Russia.

In summary, the first implication of Haraway's analysis, that we need to scrutinize the relation of self to screen, provides us with a complex set of practices and experiences. The infolding of hand and device triggers a variety of attentional states—some sub-attentional and habitual, some haptic, some fully visual, some articulating the stereo-attention of mediated elsewhere and proximate environment. These different assemblages of attention do not derive from any one technology alone (whether hardware or software) or any one particular practice; they are not reducible, in other words. Some assemblages we take up, and others sweep us up (and indeed produce us). But each assemblage produces relationships of power.

The presumption of personalized power implicit in mobile devices should not be underestimated; we can be interpellated as sovereign subjects with the world in our hands, available at the touch of a button. Through devices that connect and communicate, they also disconnect us, sort us, separate us, and control us, and we may have less agency than promised.

Camera to Object

The second implication we can draw from Haraway has to do with the materiality of the arrangement of camera to its object—which brings us to the literature of surveillance. As David Lyon (2001) maintains, surveillance is a relationship of power. Thus, in any analysis, we need to map the particularities of the situation: the location of the camera, the purpose of the camera, the presumed subject of the camera, and the organizational and institutional assemblage of which it is a part. For example, traffic cameras are part of municipal infrastructures for the management of traffic flow and density. It is one option among many to accomplish this task (from tracking the signals of mobile phones to rethinking the architecture of roadways and the structure of the urban). Tourist cameras are there to translate the location into a scene for the consumption of potential visitors (Chiba), though these can also double as devices to manage employees, customers, prevent theft, promote efficiency, and so on.

We need to consider the materiality of the camera (as Haraway considers the National Geographic cameras strapped or glued to the backs of marine wildlife) and the effects of its placement. Ike Kamphof (2011) discusses nature webcams where viewers can watch live streaming video of nests or watering holes in the wild. We need to consider the ways that the presence of this camera (and its attendant equipment) disturbs and alters the "natural" scene in front of us. For example, what happens if a camera in an eagle's nest breaks down? A tension arises between the welfare of the nest and its residents and the desires of distant viewers to watch them.

The intentions of each of the examples above (Figures 12.2–12.5) are not clear, but we may speculate on their relationship to what they survey. Figure 12.2, of Chiba, is a wide shot of the waterfront from across the water. It presents the city as landscape, as a whole, rather than focusing on a detail. Perhaps this is for the tourists' gaze. It may be sponsored by a municipality (civic promotion, harbor patrol) or private concern. But through it we grasp Chiba as an urban spectacle. The Control button on the lower right allows exploration. Figure 12.3, of the park in Seattle, is much more particular: a wide shot of a park where youth ride bikes or skateboards over and around obstacles. The shot encompasses the park in its entirety. This could be civic promotion (facilities for youth recreation in the city) or public safely (to protect the youth there, or to keep an eye on them). Notably, we are at sufficient distance that features of individuals are not distinguishable, and we are not allowed the ability to control the camera to zoom in closer on individuals (maintaining a sense of privacy and safety, especially of a vulnerable population). Figure 12.4, the street corner in Kristinehamn, is a public area of the city, but more specific than the sweep of spectacle of Chiba. We don't grasp the city as a whole but as a more intimate approach to one block. Like Figure 12.3, we are not close enough to be part of the scene, or even to identify (with) the figures there. We look down on the scene as if from a balcony, higher, it seems, than the light poles. It is surveillance from a window. Compare this image with Figure 12.2, where we look up to the city, and 12.3, where we look slightly down but are closer to being on the same level as the bike riders. Figure 12.5, the hotel reception area in Russia, is an interior scene, angled sharply so that we gaze down at the front desk. It is the ever more familiar position of the surveillance camera, anonymous and "objective." We watch over, but do not participate. Though the figures are identifiable, the camera does not allow us identification with any of them. This is the most voyeuristic of the cameras, and while it may be an attempt to appeal to potential customers, it exudes the scent of control.

Articulations, Infoldings and Ethics

What Haraway's third point suggests is that the remote cam viewers are implicated in the relations of surveillance camera and spectator. I am implicated in the organizational relations that keep Russian hotel employees under careful watch, for example, or in the close observance of Seattle's youth (to keep them out of mischief, perhaps, and to keep them from harm as well).

Kamphof argues that conservation websites leverage the phenomenological closeness (affect) of the camera image into identification with ecological causes and specific animals (usually the cute ones), which leads to donations. That is, an affective alliance is established which is then exploited for capital. Kamphof writes, "A reflex to donate in response to seeing a cute animal lacks in moral weight, even by more lenient standards" (2011, 267). In that the mobile screen is much more a part of our bodily schema than even laptop computers, the image in our hand presents an even closer intimacy, an infolding.

As David Lyon (2001) has pointed out, surveillance has two faces: care and control. Participating in this surveillance assemblage through the phone app places one within a regime of control, but not directly. You have no individual power over those surveyed (one is not a guard in the panopticon). But what is established, for oneself, is a reinforcement of the continued consciousness of surveillance in other aspects of your life, and a reinforcement of the system of cameras itself by providing to it your attention (making it a viable site for advertising, or simply through its popularity, counted in hits, encourages the site to draw on aspects of the attention economy). Following Heidegger's notion of care as an ontological closeness, a being-with, we can see that the remote locations and their subjects as infolded with one's phenomenological sense of presence and place.

Control is via the act of seeing, manipulating and holding in one's grasp the scene in question. The sight of one's fingers wrapped around the image reinforces the sense of control and domination (the world in my hand). Care is in the sense of being-with. Though the images refuse immersion, there is the being-with of the synchronic image—we move here as they are moving there (Wise 2004). Holding an image in one's hand can be a gesture of care (you're in good hands, as an insurance advertisement tells us). There is a certain intimacy to this gesture as well. As Ingrid Richardson writes, "Thus, there is a certain haptic intimacy that renders the iPhone an object of tactile and kinesthetic familiarity, a sensory knowingness of the fingers that correlates with what appears on the small screen" (2012, 144). In addition, features of the iPhone, like FaceTime, present real-time two-way video streaming. This is certainly a different genre of webcam—the private, interpersonal, video-phone call (like Skype)—which is neither public nor anonymous. These are filled with the eventfulness of the conversation, even if much of the conversation may be nonverbal, just the gentle gaze into another's face (cf. Elkins 1996). In these cases, the relationship fills the screen: we no longer seek the contingent, the proof of life, the coherence of meaning. Other apps, such as Knocking, Presence or camThis! allow one to see through the camera of another iPhone. Though similar to the FaceTime feature, these apps are less about face-to-face, than allowing others to see the scene in front of one. On the one hand, this sense of shared experience falls under care, but, on the other hand, the control aspects of surveillance, and the potential for abuse, are evident (Calvert 2000).

For Haraway, the ethical obligation she refers to connects to her work with companion animals. The "ethical obligation" is to the creature on the other end. She writes: "Specifically, we have to learn who they are in all their nonunitary otherness in order to have a conversation on the basis of carefully constructed, multisensory, compounded languages. The animals make demands on the humans and their technologies to precisely the same degree that the humans make demands on the animals" (2008, 263).

But though this is an important point that could be made of any such screen and live remote image—that it puts us in relations of care with other

humans and animals—I want to propose something at little broader or perhaps more radical, and that is taking seriously something that Jane Bennett
says about assemblage theory. Assemblage theory emphasizes the materiality of the assemblage and what she calls thing-power, a recognition of the
vitality of things. She opens her discussion with a description of some items
caught in a storm drain one morning. She writes: "I was struck as well by
the way the glove, rat, cap oscillated: at one moment garbage [that is, swept
up in a human-centered semiotic], at the next stuff that made a claim on
me" (2004, 350). The recognition of the vitalism of things, of thing-power,
is that things make claims on us.[7] She writes:

> I want to give voice to a *less specifically human* kind of materiality, to
> make manifest what I call "thing-power." I do so in order to explore
> the possibility that attentiveness to (nonhuman) things and their pow-
> ers can have a laudable effect on humans. (I am not utterly uninterested
> in humans). In particular, might, as Thoreau suggested, sensitivity to
> thing-power induce a stronger ecological sense?
>
> (Bennett 2004, 348)

To bring us to ecology, assemblage theory emphasizes the ways that
"humans are always in composition with nonhumanity" (365), and the
recognition of this composition creates a "sympathetic link" "which also
constitutes a line of flight from the anthropocentrism of everyday experience" (366).

But perhaps, in the end, this is a lot to lay on a moment of staring at a
webcam image of a Russian hotel lobby on one's iPod while waiting for
your latte at Starbucks. The challenge is to connect the webcam moment to
this sprawling, heterogeneous assemblage, especially because the trajectories of much of that which composes that assemblage don't impinge on our
perceptions.

My suggestion is modest and methodological. What the assemblage version of this moment with a mobile device suggests is the starting point for
a materialist mapping project. As Latour (2005) would put it: follow the
actors. This is quite challenging given the mobility of the actors, the ways
they keep vanishing from our conscious landscape, and their electromagnetic tethers don't impress themselves on our perception. We should not
leap too quickly to ecology, or species companionship, or common cause
with humanity. If we follow the trajectories of the actors, we get not a stable
structure, a synchronous snapshot of a network, but the sprawling convergence and divergence of series of events. The point is to help make visible
that which is continually trying to disappear in this assemblage.[8]

To return to Haraway's point about the ethical obligations of the assemblage, recognizing the kinship between people and things makes the tendency
of this new assemblage to disappear ethically problematic. I realize that
there is some contradiction in talking about technologies of visual attention

becoming invisible, but functionally they do so. As technologies vanish from attention and fall into habit (or into the walls or beneath the surface of seemingly everyday objects) they no longer make present the materiality of technologies (their thingness), and in so doing erase the reminders that technologies carry with them of their function as well as the conditions of their manufacture and reclamation. For example, with regard to function, Adam Greenfield (2006) in his overview of ambient computing points out that when technological processes become unobtrusive we don't consider their ramifications easily. His example is the MasterCard PayPass system, which enables one to just tap a special key fob (or mobile phone with Pay-Pass chip) rather than swiping a credit card. The ease of gesture allows one not to ponder concretely what is happening. As a result, PayPass users spend 25 percent more than regular credit card holders because the process has become unobtrusive. The same trend can be seen in the EZPass toll system (Leonhardt 2007). When one has to stop and hand over money at a toll booth there is popular resistance when tolls go up. One notices the difference. With the EZPass one never even slows down. A reader at the booth scans a chip on the windshield of the car and sends a bill later (or takes it out of your account). One never thinks about it. According to one study, within a decade an EZPass toll stop will be 30 percent more expensive than a cash toll.

The second issue with the disappearance of these devices is that what disappears as well is any sign which indicates their manufacturer, or even the fact of their manufacture (Schaefer and Durham 2007). A material thing had an existence before you bought it and put it in your pocket (or wall) and will have an existence after you throw it out. It seems as if our culture never learned the elementary psychological lesson of object permanence: out of sight does not mean that the object no longer exists. The tendency of these technologies to want to disappear distances us from the environmental devastation of copper strip mining practices, labor conditions in microchip factories, fossil fuel consumption and subsequent pollution of the distribution system, and the sheer toxicity of the machines as, once scrapped, they pile up in landfills, heavy metals seeping into water tables (see Gabrys 2011; Maxwell and Miller 2012). To talk of these technologies in terms of assemblage allows us to map these, draw the lines, incorporate radiation, repetitive strain injury and eye strain with heavy metals and other toxins, as well as privacy, control, regulation, convenience, productivity, profit, exploitation, dignity.

Along with function and material effects, a third ethical problem with their disappearance is that it allows us to ignore what Virilio (2000) called the *integral accident*. Every technology has inherent within it an accident particular to that technology, a disaster enfolded in its very being. For example one cannot have a plane crash without a plane, and one cannot have a plane without a (potential) plane crash. No matter how many safeguards we may establish, a nuclear accident is always somehow a part of what a nuclear power plant is. What, then, is the integral accident of the Clickable World? Data theft? Information war? Or perhaps, more likely, the disastrous inability to function that would result, should this assemblage become as much a part of our human scaffolding as it is designed to be, should it fail.

The final dimension to the ethical problem of disappearing technologies is that these are technologies—ironically—of attention (they pay attention for us), and while we forget them, they remember everything. Martin Dodge and Rob Kitchin (2007) recently brought up what they call the ethics of forgetting—that with ubicomp and new technologies and practices such as lifelogging, we need to incorporate forgetting into the assemblage.

CONCLUSION

A certain sense of agency and control is presumed by these arrangements and devices: the world is at hand. In this world, as our devices disappear into our architecture and habit, as reality is reduced to information articulated (tagged) to objects, the presumption of individual empowerment is subsumed under the broader functioning of societies of control. Attention and distraction become distributed across one's body, devices and environment. The webcam in its far-off location attends, as does the software which selects the feed, and I attend to the image, in my palm, on the go.

The hole in my hand—through it I glimpse Chiba, Kristinehamn, Greensboro—is part of the production of a new subjectivity, a new everyday, where materialities distribute agency and I am ever more implicated in ethical relations of care and control which stretch from my fingers.

NOTES

1. Previous versions of this chapter were presented at the Mobilities and Imaginaries of the Mobile Internet Workshop in Waterloo, Canada, in 2011 and the Crossroads in Cultural Studies conference in Hong Kong in 2010, and a piece was presented at a PhD course on Deleuze, ethnography, and technology at the IT University of Copenhagen in 2007. It is thus well-traveled and long-gestating. My thanks to all those who have provided such gracious feedback on it, especially Kim Sawchuk.
2. A number of scholars have begun exploring the social implications of this, from absent presence to cocooning and gating behaviors. See Bull (2007), Ito et al. (2005).
3. See also Shaun Moores's (2012) development of the idea of doubling place.
4. LiveCams, now called Live Cams Pro, is produced by Eggman Technologies and claims to have been the top paid app in the United States before Angry Birds.
5. Some cameras for this and other apps allow sound as well as image. I am focused on the visual dimensions of the experience here, but the addition of ambient sound from the remote location only enhances the transparency of the screen and the immersion in the image.

6. See, e.g., Hjorth and Richardson (2009).
7. This is a statement Bennett backs off of in her more recent treatment of this idea in *Vibrant Matter* (2009).
8. As a caveat, I agree with Grossberg (2010) here that ANT is a hyperempiricism, which has yet to theorize the conjuncture. However, I think it can be a useful way of tracing the assemblage.

SOURCES

Arns, I. "Feeding the Serpent Its Own Tail: Counterforces to Tactile Enclosure in the Age of Transparency." In *Throughout: Art and Culture Emerging with Ubiquitous Computing*, edited by Ekman, 385–398. Cambridge, MA: MIT Press, 2013.

Barthes, R. *Image-Music-Text*, translated by Heath. London: Wm. Collins, 1977.

Beller, J. *The Cinematic Mode of Production*. Hanover: Dartmouth, 2006.

Bennett, J. "The Force of Things: Steps Toward an Ecology of Matter." *Political Theory*. no. 3 (2004): 347–372.

Bennett, J. *Vibrant Matter: A Political Ecology of Things*. Durham: Duke University Press, 2009.

Bolter, J., and R. Grusin. *Remediation: Understanding New Media*. Cambridge, MA: MIT Press, 2000.

Bull, M. *Sound Moves: iPod Culture and Urban Experience*. New York: Routledge, 2007.

Calvert, C. *Voyeur Nation: Media, Privacy, and Peering in Modern Culture*. Boulder: Westview, 2000.

Cooley, H. "It's All About the Fit: The Hand, the Mobile Screenic Device and Tactile Vision." *Journal of Visual Culture*. no. 2 (2004): 133–155. doi: 10.1177/1470412904044797.

De Certeau, M., L. Giard and P. Mayol. *The Practice of Everyday Life* (vol. 2), translated by Tomasik. Minneapolis: University of Minnesota Press, 1998.

Deleuze, G. *Postscript on Control Societies. Negotiations, 1972–1990*, translated by Joughin. New York: Columbia University Press, 1995.

Deleuze, G., and F. Guattari. *Kafka: Toward a Minor Literature*, translated by Polan. Minneapolis: University of Minnesota Press, 1986.

Deleuze, G., and F. Guattari. *A Thousand Plateaus: Capitalism and Schizophrenia*, translated by Massumi. Minneapolis: University of Minnesota Press, 1987.

Dodge, M., and R. Kitchin. "'Outlines of a World Coming into Existence': Pervasive Computing and the Ethics of Forgetting." *Environment and Planning B: Planning and Design*. no. 34 (2007): 431–445. DOI:10.1068/b32041t.

Elkins, J. *The Object Stares Back: On the Nature of Seeing*. New York: Harvest, 1996.

Farman, J. *Mobile Interface Theory: Embodied Space and Locative Media*. New York: Routledge, 2012.

Foucault, M. *Discipline and Punish*. Harmondsworth: Penguin, 1977.

Gabrys, J. *Digital Rubbish: A Natural History of Electronics*. Ann Arbor: University of Michigan Press, 2011.

Gallagher, W. *Rapt: Attention and the Focused Life*. New York: Penguin Press HC, 2009.

Goggin, G. *Cell Phone Culture: Mobile Technology in Everyday Life*. New York: Routledge, 2006.

Greenfield, A. *Everyware: The Dawning Age of Ubiquitous Computing*. Berkeley: New Riders, 2006.

Grossberg, L. *Cultural Studies in the Future Tense*. Durham: Duke University Press, 2010.

Haraway, D. "Crittercam: Compounding Eyes in Naturecultures." *When Species Meet.* Minneapolis: University of Minnesota Press, 2008.

Highmore, B. *Cityscapes: Cultural Readings in the Material and Symbolic City.* New York: Palgrave Macmillan, 2005.

Highmore, B. *Ordinary Lives: Studies in the Everyday.* New York: Routledge, 2010.

Hjorth, L. "Domesticating New Media: A Discussion on Locating Mobile Media." In *Mobile technologies: From telecommunications to media*, edited by G. Goggin and L. Hjorth, 143–157. New York: Routledge, 2009.

Hjorth, L., and I. Richardson. "Playing the Waiting Game: Complicating Notions of (Tele)presence and Gendered Distraction in Casual Mobile Gaming." *Australian Journal of Communication.* no. 1 (2009): 23–35.

Ito, Mizuko, Misa Matusda and Daisuke Okabe. *Personal, Portable, Pedestrian: Mobile Phones in Japanese Life.* Cambridge, MA: MIT Press, 2005.

Jackson, M., and B. McKibben. *Distracted: The Erosion of Attention and the Coming Dark Age.* New York: Prometheus, 2008.

Kamphof, I. "Webcams to Save Nature: Online Space as Affective and Ethical Space." *Foundations of Science.* no. 2–3 (2011): 259–274. DOI 10.1007/s10699-010-9194-7.

Lanham, R. *The Economics of Attention.* Chicago: University of Chicago Press, 2006.

Latour, B. *Reassembling the Social: An Introduction to Actor-Network Theory.* Oxford: Oxford University Press, 2005.

Lefebvre, H. "Seen from the Window." *Rhythmanalysis: Space, Time, and Everyday Life*, translated by Elden and Moore. London: Continuum, 2004.

Leonhardt, D. "The Tricky Part of an E-ZPass Economy." *New York Times.* July 4, 2007.

Lyon, D. *Surveillance Society: Monitoring Everyday Life.* Buckingham: Open University Press, 2001.

Lyon, D. *Surveillance Studies: An Overview.* Malden: Polity, 2007.

Mann, S., and H. Niedzviecki. *Cyborg: Digital Destiny and Human Possibility in the Age of the Wearable Computer.* Toronto: Anchor Canada, 2001.

Marks, L. *Touch: Sensuous Theory and Multisensory Media.* Minneapolis: University of Minnesota Press, 2002.

Marks, L. "Haptic Visuality: Touching with the Eyes." *Framework.* no. 2 (2004): 79–82.

Maxwell, R., and T. Miller. *Greening the Media.* New York: Oxford University Press, 2012.

McCarthy, A. *Ambient Television.* Durham: Duke University Press, 2001.

Moores, S. *Media, Place & Mobility.* New York: Palgrave Macmillan, 2012.

Nicholson, J. "The Third Screen as Cultural Form in North America." In *The Wireless Spectrum*, edited by Crow, Longford and Sawchuck, 77–94. Toronto: University of Toronto Press, 2010.

Orgad, Shani. "This Box Was Made for Walking: How Will Mobile TV Transform Viewers' Experience and Change Advertising?" Nokia. Accessed 8 August, 2014. http://eprints.lse.ac.uk/42755/1/thisBoxWasMadeForWalkingNov2006.pdf.

Palmer, D. "Webcams: The Aesthetics of Liveness." *Like, Art Magazine.* no. 12 (2000): 16–22.

Plant, S., *On the Mobile: The Effects of Mobile Telephones on Social and Individual Life.* Chicago: Motorola, 2001.

Rheingold, H. *Smart Mobs: The Next Social Revolution.* Cambridge, MA: Perseus, 2003.

Rheingold, H. *Net Smart: How to Thrive Online.* Cambridge, MA: MIT Press, 2012.

Richardson, I. "Mobile Technosoma: Some Phenomenological Reflections on Itinerant Media Devices." *Fibreculture.* no. 6 (2005). Accessed 8 August, 2014. http://six.fibreculturejournal.org/fcj-032-mobile-technosoma-some-phenomenological-reflections-on-initerant-media-devices.

Richardson, I. "Pocket Technospaces: The Bodily Incorporation of Mobile Media." *Continuum.* no. 2 (2007): 205–215.

Richardson, I. "The Hybrid Ontology of Mobile Gaming." *Convergence.* no. 4 (2011): 419–430.

Richardson, I. "Touching the Screen: A Phenomenology of Mobile Gaming and the iPhone." In *Studying Mobile Media: Cultural Technologies, Mobile Communication, and the iPhone*, edited by Hjorth, Burgess, and Richardson, 133–151. New York: Routledge, 2012.

Richardson, I., and R. Wilken. "Haptic Vision, Footwork, Place-Making: A peripatetic Phenomenology of the Mobile Phone Pedestrians." *Second Nature.* no. 2 (2009): 22–41.

Richtel, M. "Digital Devices Deprive Brain of Needed Downtime." *The New York Times.* August 24, 2010.

Schaefer, P. D., and M. G. Durham. "On the Social Implications of Invisibility: The iMac G5 and the Effacement of the Technological Object." *Critical Studies in Media Communication.* no. 1 (2007): 39–56.

Terranova, T. "Attention, Economy and the Brain." *Culture Machine.* no. 13 (2012): 1019.

Verhoeff, N. "Grasping the Screen: Towards a Conceptualization of Touch, Mobility, and Multiplicity." In *Digital Material: Tracing New Media in Everyday Life and Technology*, edited by Boomen, Lammes, Lehmann, Raessens and Schafer, 209–222. Amsterdam: Amsterdam University Press, 2009.

Virilio, P. *The Information Bomb*, translated by Turner. New York: Verso, 2000.

Winocur, R. "Radio and Everyday Life: Uses and Meanings in the Domestic Sphere." *Television and New Media.* no. 3 (2005): 319–332.

Wise, J. *Exploring Technology and Social Space.* Thousand Oaks: Sage, 1997.

Wise, J. "'An Immense and Unexpected Field of Action': Webcams, Surveillance, and Everyday life." *Cultural Studies.* no. 2/3 (2004): 242–442.

Wise, J. "Assemblage." *Gilles Deleuze: Key concepts* (2nd ed.), edited by Stivale. Chesham: Acumen, 2011.

Wise, J. "Attention and Assemblage in the Clickable World." In *Communication Matters: Materialist Approaches to Media, Mobility and Networks*, edited by Packer and Wiley, 159–172. New York: Routledge, 2012a.

Wise, J. "Through the Looking Glass with Google." *Flow.* November 19, 2012b. Accessed 8 August, 2014. http://flowtv.org/2012/11/though-the-looking-glass/

Wise, J. "Re-Framing Google Glass." *Flow.* April 13, 2013. Accessed 18 August, 2014 http://flowtv.org/2013/04/re-framing-google-glass/.

13 Apps and Drive

Jodi Dean

In a *New York Times* article, part of a Pulitzer prize–winning investigative series on Apple, Charles Duigg and Keith Bradsher recount a dinner with Silicon Valley executives and U.S. President Barak Obama. Apparently, the President asked Steve Jobs why he didn't build iPhones in the U.S., and Jobs replied that those positions are never coming back. Explaining why, the article, also discussed in Chapter 5 of this volume, describes the scale and flexibility of the Chinese company Foxconn, the human costs of this scale and flexibility, and the effects of this scale and flexibility on U.S. ability to compete in a global market. The Foxconn facility has over two hundred thousand employees. About a quarter of them live in company barracks and work twelve-hour shifts. Such enormous capacity enables Foxconn not only to adapt to new design specifications but also to command large numbers of workers to labor overtime. Both capacities mean that the company can get new products out quickly. As an Apple supply demand manager quoted in the article says, "They could hire 3,000 people overnight . . . What U.S. plant can find 3,000 people overnight and convince them to live in dorms?" The *Times* piece recounts the erosion of mid-wage jobs in the U.S., the increase in low-wage service sector positions, and the $2 billion in stock high-level Apple employees and directors received (on top of their salaries) in 2011. Overall, the piece is a depressing venture into the bowels of production: the conditions of workers in China and the U.S. are horrible, differently so, but linked in the misery of the Apple supply chain.

Duigg and Bradsher end the article with an app. They write:

> Before Mr. Obama and Mr. Jobs said goodbye, the Apple executive pulled an iPhone from his pocket to show off a new application—a driving game—with incredibly detailed graphics. The device reflected the soft glow of the room's light. The other executives, whose combined worth exceeded $69 billion, jostled for position to glance over his shoulder. The game, everyone agreed, was wonderful.
>
> (Duigg and Bradsher 2012)

After all the inequality, the exploitative reality of the production of smartphones out of hundreds of components made around the world and assembled in China, after all the incessant executive rhetoric of competition, of can't and must, even as Apple's revenue tops $108 billion, we end up with the app, the wonderful, fascinating app capable of capturing billionaires. The app fastens all their attention to the iPhone. The device delivers a little thrill, a little wonder. The executives can hold animation in their hands and interact with it. In this moment of fascination, work and play converge; after all, they are still executives doing the work of pressuring, lobbying, and networking. Yet their fascination with the app makes them momentarily childlike. The app rivets them to the phone such that they are completely and totally present in the moment of the game, which means they are at simultaneously withdrawn from the prior context of inequality and exploitation. Fascination with the app withdraws them into immediate presence.

Apps are fascinating; they are also *fasteners*. They fasten us to our tablets and smartphones. They enthrall us, each of us, as individuals, attaching us to personalized, individualized, customized experiences of our devices. In the mashed-up metaphors of an industry fanboy celebrating the fifth birthday of the iPhone:

> The lingering bombshell that the iPhone has birthed is, needless to say, the idea of apps. Apps as a commercial prospect. Apps as a celebration of a device's identity. Apps as fun. Apps as independent spirit . . . Apple made software cool again . . . The smartest thing Apple ever did with the iPhone is open up an opportunity for excited developers to create their own love letters and tributes to the iPhone, to show Apple and everyone else what their device could become.
>
> (Stein 2012)

Apps make iPhones into affective machines. Without apps, we wouldn't need or want smartphones. Regular mobile phones would be enough for texting, leaving messages, even talking to another person. As apps fasten us to our devices, they withdraw us momentarily from our larger world, the pressures of everyday life as well as the specific sites and transactions that constitute global capitalism's technology supply chains. For some of us, actually many of us, the pressures of everyday life and the transactions that constitute supply chains are mutually interconnecting: we may work in factories, for retailers, in shipping, as developers, for corporations, as content providers and as data purveyors (insofar as our smartphones collect and transmit user and usage data). Fascination with apps is the affective attachment point tethering complex chains of production in which the app is less a product than itself another means of production. Yet with this in our hands, we aren't emancipated, controlling the conditions of our work. We are even more slaves of the machine.

This chapter considers the generative dimension of apps, their ability to tether us to devices by generating senses of attachment, enthusiasm and fascination. I attend to the ways apps amplify and consolidate specific features of communicative capitalism—network exploitation ("powerlaw" phenomena, generation of the long tail and the one), individualist and individualizing fantasies (abundance, participation, personalization) and reflexivity.[1] Before I make the argument, though, I provide some definitions and background. First, apps; then affect.

APPS

By "app" I mean a small piece of software designed for a specific narrow purpose. An app is to a software suite what a single is to an album: a short, individuated unit (Bogost 2011). It's a fragment that is no longer a part but a new, smaller whole or unit, complete by itself. My focus here is on apps for smartphones and tablets, but there are also Web apps, social networks apps, and various different kinds of apps for laptops. Although there are interesting and important differences between Web apps and mobile apps, I don't take these up but treat "app" instead as a generic term for both. I don't take a position in the debate over whether apps are killing the Web by attaching users strictly to downloaded apps. It makes more sense to me to recognize the interconnectedness of Web and phone apps, particularly given the prominence of the Facebook app (Facebook claims a share of about 20 percent of smartphone daily usage) (Lomas 2013).

Most of us associate apps with Apple, the iPhone and the slogan "There's an app for that," which was used in 2008 for the iPhone 3G and the launch of Apple's App Store. In the late nineties, though, there were already apps for mobile devices like the Palm Pilot and the handheld PCs running the Windows Mobile operating system. People could buy apps in nascent app stores like Handango.com and Pocketgear.com. Cell phones also featured a few basic apps—calendars, address books, simple games. Apple changed the playing field when it not only released the iOS software development kit so that third-party developers could write apps for the iPhone but also provided a way for developers to sell their apps. The App Store made apps available in one place. It used the payment features of iTunes to make purchasing an app quick and easy, and it made it possible to buy, download and upgrade apps nearly seamlessly with an App Store button right on the phone. Apple thus consolidated a series of previously separate actions. As it did so, it created a new market, what many tech commentators call an ecosystem so as to highlight the multiple interconnecting features of the market—developer, consumer, operating system, device manufacturer, service provider, analytics and store itself.

As of July 2013, there were over nine hundred thousand apps available in the Apple app store and there had been over fifty billion downloads. The apps sell the phone—no apps, no iPhone market. Apple's own documents are up front about this point; if developers stop developing, "customers may choose not to buy the company's products" (Streitfeld 2012). Because of apps, people have a reason to buy a smartphone; it can do a lot, more in fact, than one can even imagine.

Here are a bunch of statistics. As of December 2012, the majority of mobile phone users in the U.S. used smartphones. Ninety percent of U.S. smartphones use either Google's Android OS or Apple's iOS. As of February 2013, the majority used Android devices, although more apps are still being developed for iOS. According to an industry report, the dominance of Google and Apple is "reinforced by the two platforms' well-developed app ecosystems" (Comscore Report 2013). In the five years since the launch of Apple's App Store, there are now multiple app stores (at least seventy), the largest of which is Google Play. In July 2013, Apple abandoned a lawsuit it had been pursuing against Amazon for its app store for Android. All three companies—Apple, Google and Amazon—have app stores in multiple countries. There are also numerous country specific app stores as well as speciality and boutique app stores. Globally, as of August 2013, most app store revenue came from the U.S. (second were Japan and South Korea) (Schoger 2013). Even though Google's Android OS is on the majority of smartphones, Apple's App Store still generates more than twice as much revenue as Google Play overall (Schoger 2013). In the U.S., the top two hundred apps bring in about $1.1 million a day in Google Play. The top two hundred in the Apple App Store bring in $5.1 million a day (van Agten 2013).

In the U.S., four out of five mobile minutes are spent engaging with apps. One in three minutes spent online happens via phones and tablets—people aren't tied to their laptops; they are tied to, nearly inseparable from, their phones. A recent report finds that over seventy percent of U.S. mobile users have their phones with them twenty-two hours a day. They check their phones as soon as they wake up and then over a hundred more times until they go to sleep (Stern 2013).

Affect

Contemporary communications media capture users in intensive and extensive networks of enjoyment, production and surveillance. My term for this formation is communicative capitalism.[2] Just as industrial capitalism relied on the exploitation of labor, so does communicative capitalism rely on the exploitation of communication. Communicative capitalism is that economic-ideological form wherein reflexivity captures creativity and resistance so as to enrich the few as it placates and diverts the many. The media practices we enjoy, that enable us to express ourselves and connect with others, appropriate and reassemble our longings into new forms of exploitation and control.

My thinking about communicative capitalism draws from psychoanalysis, particularly from Jacques Lacan as interpreted by Slavoj Zizek. Lacan's version of the psychoanalytic concept of drive is particularly useful because it provides a way to understand the reflexive structure of complex networks.

Lacan conceives drive as necessarily death drive (unlike Freud, who presents *eros* as also a drive). Rather than involving some kind of primary homeostasis or equilibrium, drives are a destabilizing force, a force that persists, that exerts a pressure, without regard for the pleasure and well-being of the subject. As a persistence that is not for the sake of the subject, drive has almost an un-dead quality. It is excessive, beyond the ostensibly natural contours of life and death. In the Lacanian view, then, drive as death drive encompasses the way that even a drive for life results in paradoxes wherein saving life entails sacrificing it, pursuing life leads to risking it, and cherishing life looks like a bizarre fixation on morbidity. Drive is negativity as a force.

Drive is also reflexive. It turns back in on itself, even into its opposite. Freud's example is of how voyeurism becomes masochism. Why is it reflexive or why is drive a turning back? The answer is because drive, in psychoanalysis, is the force loss exerts on the field of desire. The familiar Freudian term here is "sublimation." That the drive is thwarted or sublimated means that it reaches its goal by other means, through other objects. Blocked in one direction, it splits into multiple vectors, into a network. If Freud views the process as akin to the flow of water into multiple tributaries and canals, we might also think of it as an acephalic power's attempt to constitute and reach its objects by any means necessary—and then to do it again and again and again and again, getting a little enjoyment in each repetition. Highlighting this reflexive dimension of drive, Lacan describes it as a loop, like that of boomerang. I think of it as like a slot machine or like drifting through Facebook or even fiddling around with the apps on my phone, which is sort of fun because they make the phone look cool even if I am not really doing much of anything. It's a kind of mindless repetition that fills the gaps.

The best example of drive as a reflexive, negative force comes from the comedian Louis C.K., in an appearance on the Conan O'Brien show, the video of which has over five million views on YouTube (Team Coco 2013). Louis C.K. describes how phones take away the ability to just sit there. He explains this in a perfectly Lacanian fashion: underneath the person is the Thing, the forever empty, the knowledge that it is all for nothing and you're alone. Our phones let us avoid confronting this horror at the basis of being a person—which is why, Louis C.K. says, that 100 percent of people driving are texting; they will risk killing themselves or another person just to avoid confronting this terrifying abyss. The phone lets us be neither completely sad nor completely happy, just kind of satisfied. He thus beautifully explains the basic Lacanian point that whereas with desire we seek an enjoyment we can never attain, with drive we end up with a little stain of enjoyment we can't eliminate.

Drives are also partial. Lacan specifies this idea as "partial with regard to the biological finality of sexuality" (Lacan 1998). I understand the point to refer to the variety of changing, incomplete, and dispersed ways subjects enjoy. Drives do not develop in a linear fashion from infant to adult. Rather, they fragment and disperse as they satisfy themselves via a variety of objects. As Joan Copjec writes, "It is as if the very function of the drive were this continuous opening up of small fractures between things" (Copjec 2002, 43). Her language here is precise: the fractures are not of things but between them; the parts that are objects of the drives are not parts of wholes but parts that appear in the force of loss as new expressions of a whole (she uses Gilles Deleuze's example of the role of the close-up as a cinematic device: it's not part of a scene enlarged; rather, it's an expression of the whole of the scene) (Copjec 2002, 53). Lacan refers to the partial object as an object of lack, an object that emerges in the void of the drive to provide the subject with satisfaction.

Correlative to the part is a further aspect of drive that Lacan renders as montage, a constant jumping without transition between heterogeneous elements. Montage suggests movement without message, movement with intensity, movement outward and back. Disparate images and sounds shift and mutate without beginning or end, head or tail. Lacan: "I think the resulting image would show the working of a dynamo connected up to a gas-tap, a peacock's feather emerges, and tickles the belly of a pretty woman, who is just lying there looking beautiful" (Lacan 1998, 169). More contemporary ways to understand montage are mash-ups, samples and remixes—or, better, our engagement with the media, devices and networks of communicative capitalism. We check messages, snap and share selfies, play games, favorite a tweet, again and again. We close one app and open another, fingers sliding quickly back and forth across our touch screens in gentle repetitive strokes. Drive circulates, round and round, producing satisfaction even as it misses its goal.

I've emphasized the negativity of drive as well as drive's fragmentation, partiality and circulation. One last general point regarding my approach to affect via psychoanalysis and the drive: as Joan Copjec argues, Freud and Lacan associate affect with movement. Freud views affect as displacement. Conventional readings of Freud construe this displacement as the way intense feeling distorts perception. Copjec disagrees, arguing that the affective experience of something as moving indexes a movement beyond the perceiving individual, a surfeit or excess that ruptures the perception, making it more than itself and enabling it to open up another register (for Lacan, the Real; for Deleuze, the virtual). Affect is this movement that estranges the subject from its experience. A thought, memory or perception is affective to the extent that it opens up or indexes something beyond me. The dimension of affect is this "more than a feeling" that imparts movement. At the same time that affect is movement, moreover, there is also a specific affect that is a halting or arrest. Copjec invokes the image of running in place. This affect that is an inhibition of movement is anxiety. For Lacan, the object of anxiety is surplus *jouissance, objet petit a* (Lacan 2007, 147). The experience of anxiety is a confrontation with excessive enjoyment, a paralyzing confrontation wherein one encounters what is in one more than oneself, an alien yet intimate kernel at the core of one's being—Louis C.K.'s forever empty. Copjec writes, "Jouissance makes me me, while preventing me from knowing who I am" (Copjec 2002, 102). Finding oneself face-to-face with *jouissance* or thrown into awareness of one's own inhuman double, one is pulled between incomprehensibility and extreme intimacy.

Such a transfixed pull is another word for fascination. It would seem, then, that fascination is an affect like anxiety. It arrests via a confrontation with *objet a* or a little nugget of *jouissance*. The attempt to escape anxiety results in capture at another point or level; this capture is the immobility of fascination. One is fastened to a thing or activity. In *Seminar XI*, Lacan notes that in his account of scopic drive Freud is able to link seeing and being seen in the fact that the subject sees himself (Lacan 1998, 194). Drive involves making oneself seen, as well as, Lacan adds, making oneself heard, and in so doing making oneself (Lacan 1998, 195). Apps fasten us to our personal devices as they let us see and hear ourselves. The difference between smartphones and regular mobile phones is that the latter are for communicating with others. The former are about making ourselves: I see myself on the Temple Run; I see myself in the alignment of three matching candies; I see myself being seen by others as I ask them for help on a level or share with them my scores or post my scores in social media; I see myself in my busy calendar, seeing how important I am to others, my indispensability to work and friends; I see others seeing me as I quickly snap and share selfies, more concerned with appearing to them than with them appearing to me. Most of the time, or at least too easily, we *enjoy* the fact of making ourselves seen rather than think much about others' reactions to or thoughts about the various instances of our self-making. It should come as no surprise that taking photos is second only to texting as the most frequent smartphone user activity (Comscore Report 2013, 33). Apps let us quickly engage others as the figures in our fantasies who watch us, anywhere, anytime. We imagine them seeing us and we feel important, connected, excited.

In sum, psychoanalysis opens up the way that people enjoy the circulation of affect generated through reflexive communication. The very system—the movement from my immediacy, to my viewing of my immediacy on my smartphone, to my sharing of my now mediated immediacy in ways that will immediately exceed my control—is captivating, generating and circulating affect as a binding technique. Bound to their phones, people are tethered to the network which means, as an industry report observes, "more monetization opportunities for media companies" (Comscore Report 2013, 11).

A behavioral study of the habits of smartphone users highlights, in a different vocabulary, some of the ways apps fasten users to phones. Seeking quick access to rewards, users quickly and frequently check various applications. Their interaction with their phones is characterized by short duration, isolated, reward-based sessions that respond easily to cues. Checking behavior can be stimulated and made to increase with apps that provide more rewards (Oulasvirta 2011).[3]

Apps are fabulous inventions for the circulation of drive. To enjoy the mindless pleasure of moving through digital environments we don't have to be stuck at our laptops. Enjoyment is in our hands. The language mediating our engagement with smartphones exudes this enjoyment: customers are depicted as fans, as enthusiasts, as addicts, as obsessed. They love their phones, can't live without them, are constantly stroking them and attending to them. With the right apps, phones and tablets soothe our children and help us lose weight. They can be and do anything we want, promising to bring us what we desire even as they function more as an object-cause entrapping us in drive.

I turn now to some of the specific ways that apps reinforce dominant elements of communicative capitalism. I begin with exploitation in complex networks.

NETWORK EXPLOITATION AND POWERLAWS

Since Apple released the iOS SDK and opened the App Store, apps have been championed as the next big thing in the digital economy. Industry reports present apps as leaders in innovation responsible for significant job creation—hundreds of thousands of new jobs in the U.S. and UK (Mandel 2012; "European App Economy" 2013) are ostensibly so generative that they don't just stimulate job growth; they incite entire app ecosystems that build around a basic platform and include developers, device manufacturers, service providers, and advertisers. Congratulating itself as a job creator, Apple takes credit for generating more jobs for UPS drivers (Streitfeld 2012).

Initially, apps were celebrated as opening up opportunities for independent and small-scale developers to break into the software market. Now, apps are recognizably elements in an intensely competitive market with high entry costs. With nearly a millions apps, getting attention is so difficult that there are now companies that specialize in marketing apps, in hosting conferences and workshops on app marketing, and in selling app analytics that can help developers identify usage patterns and better tailor their apps to the markets they want to reach. It's no wonder that already existing and well-known companies have a strong advantage. Celebration of and competition within the app market are connected: the opening of the market to independent developers generated interest in and attachment to the devices, thereby further empowering the powerful without costing them anything. Apple didn't have to pay its own developers—it could benefit from the free labor of thousands and even charge them a fee for being including in the App Store. Essentially, Apple offloads development costs and risks onto digital pieceworkers while it reaps the rewards.

Smartphone users typically have around thirty apps on their phones—there are limits to how many apps a given consumer can use (Google Report 2012). There are hundreds of thousands of apps, billions of downloads, but only a handful are widely popular. Here are some examples: the top free apps in the Apple App Store in August 2013 were the games Despicable Me: Minion Rush, Candy Crush Saga, Plants v. Zombies 2, YouTube and Google Maps. The top free apps on Google Play were Facebook, WhatsApp Messenger, Candy Crush Saga, Line: Free Calls and Messages, and Facebook Messenger (Schoger 2013). The top publishers of free apps in both stores were Google, Facebook, Apple and Disney. Google, Apple and Disney were also in the top ten publishers of paid apps, which also included game publishers like Gameloft and Electronic Arts.

The same pattern of the dominance of the few and the weakness of the many reappears in a survey of 270 North American app developers. The vast majority make less than $15,000 a year. A very few make over $100,000 (Cravens 2012). This pattern of few winners/many losers dominates networked interactions. Complex networks are characterized by free choice, growth, and preferential attachment. As Albert-Laszlo Barabasi demonstrates, complex networks follow a powerlaw distribution of links (we can also say hits, downloads, views, purchases, etc.) (Barabasi 2003). The item in first place or at the top of a given network has twice as many links as the item in second place, which has more than the one in third and so on, such that there is very little difference among those at the bottom but massive differences between top and bottom. So lots of novels are written. Few are published. Fewer are sold. A very few become best sellers. Or lots of articles are written. Few are read. The same four are cited by everybody. The idea appears in popular media as the 80/20 rule, the winner-take-all or winner-take-most character of the new economy, and the "long tail."

Exploitation in communicative capitalism consists in stimulating the creative production of a broad field in the interest of finding and monetizing the one. Expanding the field produces the one (or, hubs are an immanent property of complex networks). Instead of working for pay, one works for a chance at pay. Instead of Apple paying large teams of app developers, hordes of independent developers work for free, hoping that their apps will make something. Most won't break even.

INDIVIDUALIST AND INDIVIDUALIZING FANTASIES

The fantasy, though, is that every app will be another Instagram or Angry Birds. This leads me to a second way that apps reinforce communicative capitalism—fantasy. In *Democracy and Other Neoliberal Fantasies*, I present three animating fantasies in communicative capitalism: abundance, participation and wholeness. Apps extend all three—the sense that every-thing is out there and available (there's an app for that), the sense that one is active and connected, *really involved*, when one is on call all the time, and the sense that one is part of a whole, integrated into a world and that the world accessed through the screen is that world. At the same time, however, apps amplify an additional fantasy, that of the individual as a unique locus of meaning and action. Apps are individualist and individualizing.

Apps intensify the fantasy of the individual at the level of the developer as well as that of the user. First, with regard to developers: in the five years since Apple released the iOS SDK, apps have been celebrated as programs that individuals can write. Unlike large software projects, the specialized focus of apps are within reach of individual developers, particularly developers laid off in the ongoing recession. App development could thus initially be imag-ined as the way an individual could hit it big, make millions, and do it inde-pendently. The app fantasy combines independence, creativity and financial success. In the words of one industry commentator, "The free-market feel of an app store suddenly allowed masses of creative free thinkers to suddenly dream of making millions nearly overnight" (Stein 2012). The app developer fantasy individualizes the developer as a creative entrepreneur, going it alone and beating the odds. His freedom is unleashed by the app store, itself a per-fect, shining example of all that is good about free markets.

Although the dominance of corporations like Google, Facebook and Dis-ney in the app marketplace has eroded the plausibility of the individualist fantasy, it hasn't eliminated it completely. Instead, the fantasy is itself mobi-lized by analytics companies trying to attract developers to their services and to convince them that, with the proper tools, they can beat the odds. Analytics let the developers see themselves being seen—they can see how their users, customers and clients see them, how they engage them. The interaction of consumer and app isn't hidden; with the proper analytics, developers can see it and thus themselves.

The fantasy of the individual developer is also mobilized by lobbyists try-ing to influence regulatory regimes. Relying on the image of the individual striving to be free to be creative, industry groups argue for the construction and regulation of markets freed from previous guarantees to workers. They present apps as a technological fix, a remedy to ailing economies, even as they use apps as opportunities to create new markets. For example, the group ACT, the Association for Competitive Technology, recently sponsored a report on the European App Economy. ACT presents itself as an international advocacy group for small mobile app developers and "other small business innova-tors." Its goals are to "help its members leverage their intellectual assets to raise capital, create jobs, and continue innovating." But its "sponsor mem-bers" include Apple, AT&T, BlackBerry, Facebook, Intel, Microsoft, Oracle, PayPal and Verizon.

As part of its emphasis on innovation, ACT opposed the U.S. Federal Trade Commission's suit against Intel. It endeavors to strengthen intellectual property, copyright, and patent law (in the U.S. and EU). As ACT4Apps it has spearheaded a privacy initiative—this subset of ACT (recall, a small business group) is comprised of Facebook, BlackBerry, Apple and the data privacy management company TRUSTe. Presenting itself as supporting innovators, ACT pushes the fantasy of wholeness: it declares on its website, under the heading, *One World. One Agenda:* "Regardless of region or nationality, small business innovators have largely the same interests from governments and regulators: access, flexibility, and consistency." Within this vision of wholeness–clearly ideological—it celebrates the individual innovator as an entrepreneur, and it uses this vision in its lobbying pressure: the report it sponsored on the European App Economy claims that apps are driving productivity, growth and jobs and that governments need to, indeed, must support this economy with a supportive and flexible business environment, releasing more spectrum for wireless services, making more government data available to app developers and embracing app-driven development across all sectors. The potential incongruity here—if the app economy is already creating jobs and wealth why is more support necessary?—is pushed aside by transformation-innovation tech-hype ("mobile apps represent the next phase of the ICT revolution"). What is presented as the transformative potential of apps—all the innovations and enhancements to every part of life that apps provide—becomes itself dependent on, an argument for, specific legal and regulatory transformations. The report says that the benefits of ICT required markets to be transformed; "organizations and individuals needed to find different ways of doing things" ("European App Economy" 2013, 4). For transformation to occur likewise in the app economy, labor and product markets will have to undergo further flexibilization. One such flexibilization, associated with apps and enterprise services, is "bring your own device," which means that workers are to supply their own instruments of production. The benefit of this, according to ACT, is that the cost of ownership is diminished. We should see it as part of the ongoing efforts to fragment workers—which ACT explicitly encourages as it presents "hobbyists and part-timers" as key to "the job creation process in the App Economy" ("European App Economy" 2013, 10)

An individualist fantasy promotes app development. App consumption occasions a corresponding individualization of consumers. There are at least four components of this individualization all of which contribute to the enclosure of collective desires into individuated persons, dispersion of collective subjects and exacerbation of the fantasy of individual separateness and omnipotence.

1. *Isolation.* This first component is the separation of the person from her setting and her reconnection to another one, one that is familiar to her as well as traceable and available for further interruption via prompts and suggestions. Instead of present to her surroundings, to the people around her, to a broader context that is changing, uncertain

and out of her control, apps fasten an individual to a device that is just for her, that makes her the center of a world that she can open and close, enter and leave, that provides her with attention, little bits of enjoyment and a kind of enjoyment that is hers alone. She doesn't have to share it with those around her.

2. *Personalization.* This second component is just the latest in the broader trend in computing to attach individuated persons to their own devices. Apps amplify this insofar as they make separate operations available for a more customized device experience. Each person's phone is a little bit different. The ease of specification cultivates a sense of uniqueness—as was clear in the marketing of iPhone 5. This uniqueness, though, is fragile and imaginary, increasing people's anxiety—are *they really all that unique?*—and inducing them to reassure themselves by stroking and engaging their phones. This personalization is amped up in the new emphasis on education apps (particularly for iPads). Using a bond issue (that is, debt) the Los Angeles unified school district has worked to give each of its students an iPad. In addition to being good for Apple, the iPad lets students learn individually; their learning experience is completely customized. The personalization of education means that rather than seeing themselves as members of a class, students see themselves as individuals with their own unique needs and talents that must be cultivated and addressed. It also means that education as a collective responsibility fades before the conviction that education can only ever be individual and thus the individual's own singular responsibility.

3. *Responsibilization.* The third component of individualization, responsibilization, has been a primary feature of the rapid take-up of smartphones for 24/7 work. People feel like they need their phones in order to keep up email, be available when needed, etc. This is the well-documented expansion of the working day such that some contemporary workers are never not at work. In its newer incarnation, apps are associated with the individuation of responsibility. They enable the downloading of responsibility from organizations onto individuals. Customers are responsible for tracking their flight status; it's not the obligation of the airline. Rather than illness being a medical concern that one confronts by going to a doctor, health and well-being are the salient factors for which one must accountable. There is growing investment in a wide array of wellness apps—the digital health ecosystem—that encourage health measurement and monitoring, the self-management of medical conditions via reminders and logs, as well as access to medical, billing and insurance records ("European App Economy" 2013, 20). This responsibilization affirms the individual fantasy of omniscience and autonomy: *I don't need anyone else; I can do it all by myself.* It does so by covering over the fact that under capitalism the matters most significant to our lives—work, education,

health, housing—are not under our control at all, are in fact determined by market forces directed toward the benefit of the few and the exploitation of the many.

4. *Identification.* An analyst at Atboy writes, "The best apps use location data, behavioral triggers, historical usage patterns and other data to create a very relevant, contextualized and personal experience" (Josh 2013). The best apps, in other words, fully locate an individual in time, place and social networks. "Best" here means most popular as well as most profitable. The more identifiable an individual's usage habits are, the more "opportunities to monetize incremental consumption activity" (the language is from an industry report). This monetization involves not just ads or upgraded versions of simple apps. It includes in-app purchasing options—for example, the game Candy Crush Saga lets players buy boosts and extra moves that will let them get through difficult game levels. A further twist here: identification can also be self-identification such that it not just apps enabling us to be identified, but our own enjoyment of apps because of the ways they let us identify ourselves. In games, this happens through leaderboards and play with other people. It also happens in that wide array of apps that encourage us to make ourselves visible to others by registering our location. Although this seems an instance of the extreme fragility of the imaginary individual, one that is competitive and insecure and always suspicious that more interesting people are doing more interesting things elsewhere, it might also be an example of the inward turn of drive. We signal where we are not because we really want others to join us but because we see ourselves where we are through them.

REFLEXIVITY

In communicative capitalism, the loop of reflexivity is a form of absorption and capture—you might think of vicious cycles here. We saw an intensive version of such reflexivity in the finance sector with the crisis of 2008: the very measures taken to manage risk were the ones that increased it. Reflexivity with respect to apps is already part of the app ideology, of the self-understanding of the industry: prominent in its current rhetoric is the term "app ecosystem." For example, some cellular service providers are considering whether they should also set up app stores, creating viable ecosystems where they can better dictate specifications that will suit their interests, interests that are primarily in apps that "drive up mobile data usage and generate new revenues" (Dalgety n.d.). Cellular networks want users to need and use more bandwidth, so they want more of the right kinds of apps in the hands of their customers.

A further example of reflexivity in the app ecosystem is a project of Facebook and Cisco Systems. They are offering free Wi-Fi to users who use Facebook to check in at certain locations, like specific restaurants and businesses. Checking in through Facebook, the customer accesses the business's Wi-Fi network, providing Facebook and the business with data like age, gender, city, friends. Cisco says that helps provide "a very personalized mobile experience for customers while they shop, spend time with friends and family." It also provides businesses with "more opportunities to connect with their customers," that is, with data that will enable them better to target their customers with ads and promotions (Albanesius 2013).

The reflexivity of the app ecosystem appears like the self-conscious neoliberalization of software, one that fragments and stimulates in order to recombine under market conditions conducive to the generation of the one, that is, to furthering the dominance of, the capital accumulation of, the already prominent. Development work is piecemeal, part-time, done and only sometimes remunerated. Centralized in stores, access to which developers have to pay—whether or not they are hobbyists, whether or not the app is free—apps are part of a market that structures reward according to powerlaws, where the winners take all and there is a long tail of losers. Yet to further this economy, to create an environment in which this sort of app ecosystem will flourish, legislators and policy-makers are encouraged to further erode guarantees to workers in favor of guarantees to property—the system turns back in on itself to retroactively generate its own conditions.

And one of the primary conditions, the one that keeps the system running, that keeps up demand for smartphones, which would be nothing without apps, is fascinating the user. App analytics search for new ways to fasten the consumer to apps, whether these are health, entertainment, personal work organization or social networking (all of which together and singly rely on the loop of drive such that they shift between seeing to being seen to making oneself seen). An industry analyst focusing on the widespread availability of analytics notes that

> Flurry Analytics has sections that report on session duration, session frequency and overall rate of retention as an application ages. Developers can customize how they collect data through events tracking . . . such as when someone likes a status, shares an article, beats a level or makes a purchase. Once events are set up, developers can also segment out sections of their audience according to behavior (i.e., purchasers) or according to more traditional audience metrics like age, gender or location (Josh 2013).

App use is traced and stored. It is use plus the record of use. The record of use is parsed and analyzed, new knowledge which is then put back to work in order to generate apps that are more fascinating, and users enjoy it. Critics have been highlighting what they see as the failure of young people to appreciate privacy, their seeming lack of concern with surveillance. I think this worry is misplaced. More significant is the way that the app ecosystem captures us in circuits where we enjoy making ourselves seen.

NOTES

1. For more detailed discussion of network exploitation, see Jodi Dean, *The Communist Horizon* (London: Verso, 2012); for more detailed discussion of fantasy in communicative capitalism, see Jodi Dean, *Democracy and Other Neoliberal Fantasies* (Durham: Duke University Press, 2009); for more detailed discussion of reflexivity, see Jodi Dean, *Blog Theory* (Cambridge, UK: Polity, 2010).
2. See Jodi Dean *Publicity's Secret* (Ithaca: Cornell University Press, 2002) as well as the books listed in note 1.
3. See also Elizabeth Cohen, "Do You Obsessively Check Your Smartphone?" *CNN Health*, July 28, 2011.

SOURCES

Albanesius, C. "Facebook, Cisco Offering Free Wi-Fi for Check-ins." *PC*. October 3, 2013. www.pcmag.com/article2/0,2817,2425179,00.asp

Barabasi, A. *Linked*. New York: Plume, 2003.

Bogost, I. "What Is an App? A Shortened, Slang Application." January 12, 2011. www.bogost.com/blog/what_is_an_app.shtml

Comscore Report, "Mobile Future in Focus 2013." February 25, 2013. www.comscore.com/Insights/Press_Releases/2013/2/comScore_Releases_the_2013_Mobile_Future_in_Focus_Report

Copjec, J. *Imagine There's No Woman*. Cambridge, MA: MIT Press, 2002.

Cravens, A. "A Demographic and Business Model Analysis of Today's App Developer." *Report for Gigacom Pro*, September 2012. http://appdevelopersalliance.org/files/pages/GigaOMApplicationDevelopers.pdf

Dalgety, R. "Building a Mobile App Ecosystem to Drive Mobile Usage." *Telecoms*. www.telecoms.com/11685/building-a-mobile-app-ecosystem-to-drive-mobile-usage/

Duigg, C., and K. Bradsher. "How the US Lost Out on iPhone Work." *The New York Times*. January 21, 2012.

"European App Economy: Creating Jobs and Driving Growth." *Report for Vision Mobile*. September 2013. www.visionmobile.com/product/the-european-app-economy/

Google Report, "Our Mobile Planet: United States of America." May 2012. http://services.google.com/fh/files/misc/omp-2013-us-en.pdf

Josh. "App Analytics, or the death of the independent app developer." *XDAdevelopers*. May 29, 2013. www.xda-developers.com/android/app-analytics/

Lacan, J. *The Four Fundamental Concepts of Psychoanalysis, The Seminar of Jacques Lacan, Book XI*. New York: Norton, 1998.

Lacan, J. *The Other Side of Psychoanalysis, The Seminar of Jacques Lacan, Book XVII*. New York: Norton, 2007.

Lomas, N. "The App Economy Is in Rude Health, Says Flurry, but Mobile Browsers Are Being Squeezed by Facebook." *Tech Crunch*. April 3, 2013. http://techcrunch.com/2013/04/03/apps-vs-mobile-web/

Mandel, M. "Where the Jobs Are: The App Economy." *TechNet*. February 7, 2012. www.technet.org/wp-content/uploads/2012/02/TechNet-App-Economy-Jobs-Study.pdf

Oulasvirta, A., T. Rattenbury, M. Lingyi and E. Raita. "Habits Make Smartphone Use More Pervasive." *Pers Ubiquit Comput* (2011). DOI 10.1007/s00779–011–0412–2. http://www.hiit.fi/u/oulasvir/scipubs/Oulasvirta_2011_PUC_HabitsMakeSmartphoneUseMorePervasive.pdf

Schoger, C. "Top Global Apps." *Distimo*. September 19, 2013. www.distimo.com/blog/2013_09_top-global-apps-august-2013/

Stein, S. "Five Years in, the iPhone's Greatest Legacy: Its Apps." *Cnet*. June 28, 2012. http://news.cnet.com/8301–13579_3–57462881–37/five-years-in-the-iphones-greatest-legacy-its-apps/

Stern, J. "Cellphone Users Check Phones 150x/Day and Other Internet Fun Facts." *ABC News*. May 29, 2013. http://abcnews.go.com/blogs/technology/2013/05/cellphone-users-check-phones-150xday-and-other-internet-fun-facts/

Streitfeld, D. "A Boom Lures App Creators, the Tough Part is Making a Living." *The New York Times*. November 18, 2012.

Team Coco. "Louis C.K. Hates Cell Phones." YouTube. September 20, 2013. Accessed March 24, 2014. www.youtube.com/watch?v = 5HbYScltf1c

van Agten, T. "A Granular App Level Look at Revenues: Google Play vs. Apple App Store." *Distimo*. May 29, 2013. www.distimo.com/blog/2013_05_a-granular-app-level-look-at-revenues-google-play-vs-apple-app-store/

Contributors

Darin Barney is Canada Research Chair in Technology & Citizenship, and an associate professor at McGill University. He is the author of *Communication Technology: The Canadian Democratic Audit* (2005); *The Network Society* (2004); and *Prometheus Wired: The Hope for Democracy in the Age of Network Technology* (2000). His current work focuses on the politics of resource infrastructure in Canada, including projects on the transformation of grain-handling technology in the Canadian prairies and the politics of petroleum and gas pipelines in the Pacific Northwest.

Enda Brophy is an assistant professor in the School of Communication at Simon Fraser University. His work has appeared in the *Canadian Journal of Communication*, *Journal of Communication Inquiry*, *ephemera*, and *Work Organisation, Labour and Globalisation*. He is the translator of *The Production of Living Knowledge: Crisis of the University and the Transformation of Labour in Europe and North America*, by Gigi Roggero.

Barbara Crow is Dean, Graduate Studies and Associate Vice President, Graduate at York University. With Professors Michael Longford and Kim Sawchuk, they cofounded *wi: journal of mobile media* and the Mobile Media Lab co-located at York and Concordia Universities. They have also co-edited *The Wireless Spectrum: The Politics, Practices and Poetics of Mobile Communication*, University of Toronto Press, 2010. Her most recent work is on aging, communication and mobilities, *http://a-c-m.ca*.

Greig de Peuter is an assistant professor in the Department of Communication Studies at Wilfrid Laurier University. He is co-author, with Nick Dyer-Witheford, of *Games of Empire: Global Capitalism and Video Games*, and, with Stephen Kline and Nick Dyer-Witheford, of *Digital Play: The Interaction of Technology, Culture, and Marketing*.

Jodi Dean is the Donald R. Harter '39 Professor of the Humanities and Social Sciences at Hobart and William Smith Colleges in Geneva, New York. She is the author or editor of eleven books, including *Democracy* *and Other Neoliberal Fantasies* (Duke 2009), *Blog Theory* (Polity 2010), and *The Communist Horizon* (Verso 2012).

Jason Farman is an assistant professor at University of Maryland, College Park in the Department of American Studies, the Director of the Design I Cultures + Creativity Program, and a faculty member with the Human-Computer Interaction Lab. He is author of the book *Mobile Interface Theory: Embodied Space and Locative Media* (Routledge, 2012—winner of the 2012 Book of the Year Award from the Association of Internet Researchers). His second book is an edited collection titled *The Mobile Story: Narrative Practices with Locative Technologies* (Routledge Press, 2014). He has published scholarly articles on such topics as mobile technologies, Google maps, social media, videogames, digital storytelling, digital performance art and surveillance. He received his PhD in Performance Studies and Digital Media from the University of California, Los Angeles.

Gerard Goggin is ARC Future Fellow and professor of Media and Communications at the University of Sydney. His research interests lie in mobiles, Internet and new media; media policy and regulation; and disability. Gerard's books include *Routledge Companion to Mobile Media* (2014; with Larissa Hjorth); *Disability and the Media* (2014; with Katie Ellis); *New Technologies and the Media* (2011); *Global Mobile Media* (2011), *Cell Phone Culture* (2006), and, with Christopher Newell, *Disability in Australia* (2005) and *Digital Disability* (2003). His edited collections on mobiles include *Locative Media Mobile* (2015) and *Mobile Technology and Place* (2012), both with Rowan Wilken, *Mobile Technologies: From Telecommunications to Media* (2009; with Larissa Hjorth) and *Mobile Phone Cultures* (2008). Gerard has also published three edited volumes on the Internet: *Routledge Companion to Comparative Internet Histories* (2016), *Internationalizing Internet: Beyond Anglophone Paradigms* (2009) and *Virtual Nation: The Internet in Australia* (2004).

Jan Hadlaw is an associate professor at York University in Toronto, Canada. Her research and teaching interests include media history, science and technology studies, and the role of design in everyday life. Her current research examines the role played by technology and design in the construction of Canadian national identity. Her work has appeared in *Space and Culture, Design Issues*, and *Objets et communication, MEI (Médiation et information)*.

Andrew Herman is an associate professor of Communication Studies in the Faculty of Arts on the Waterloo campus. He received his BA in Government from Georgetown University and his PhD in Sociology from Boston College. Before joining Laurier in 2004, he taught at Drake University, Boston College, College of the Holy Cross, and York University. He has been chair of the Communication Studies department and Director of the MA

Program in Cultural Analysis and Social Theory. He has written widely in the field of social theory, media and culture and his appeared in scholarly journals such as *Cultural Studies, Critical Studies in Media Communication, South Atlantic Quarterly,* and *Anthropological Quarterly.* Among his many publications are *The "Better Angel" of Capitalism: Rhetoric, Narrative and Moral Identity Among Men of the American Upper Class* (Westview, 1999) and his edited collections, *Mapping the Beat: Popular Music and Contemporary Cultural Theory* (Blackwell, 1997) and *The World Wide Web and Contemporary Cultural Theory* (Routledge, 2000).

Rich Ling (PhD, University of Colorado, sociology) is the Shaw Foundation Professor of Media Technology, Wee Kim Wee School of Communication and Information, Nanyang Technological University, Singapore; he is affiliated with Telenor Research and has an adjunct position at the University of Michigan. Ling has studied the social consequences of mobile communication for the past two decades. He has written *The Mobile Connection* (Morgan Kaufmann, 2004), *New Tech, New Ties* (MIT Press, 2008) and most recently *Taken for Grantedness* (MIT Press, 2012). He is a founding co-editor of *Mobile Media and Communication* (Sage) and the Oxford University Press series *Studies in Mobile Communication.*

Vincent Manzerolle (PhD) is a lecturer in the Faculty of Information and Media Studies at the University of Western Ontario. He has published articles in *ephemera, Surveillance and Society,* and *TripleC: Communication, Capitalism, & Critique,* and is the co-editor of *The Audience Commodity in a Digital Age* (published by Peter Lang).

Alison Powell is assistant professor in Media and Communications at the London School of Economics. Her research analyzes the expansion of open-source cultures and modes of production, the design and politics of the development of new information communication and communication technologies (ICTs), and the processes of Internet policy formation. She is a member of the European Network of Excellence on Internet Science where she investigates how civil society and entrepreneurs negotiate values and standards for the future Internet, and co-leads the work package on standardization and legislation activities. Previous research projects have been funded by the Canadian Social Sciences and Humanities Research Council, the Oak Foundation, the UK's Engineering and Physical Science Research Council and the National Centre for Research Methods.

Kim Sawchuk is a professor in the Department of Communication Studies at Concordia University in Montreal, QC, Canada and Concordia University Research Chair in Mobile Media Studies. She co-directs the Mobile Media Lab-Montreal with Owen Chapman (www.mobilities.ca). Her current research explores the intersection of mobility studies and critical disability studies as well as aging in a digital world.

Thom Swiss is a professor of Culture and Teaching at the University of Minnesota. Author of two books of poems, *Measure* and *Rough Cut*, he is the editor or co-editor of books on popular music, including *Bob Dylan: Highway 61 Revisited* (University of Minnesota Press, 2009), as well as books on new media literature and culture, including *New Media Poetics: Contexts, Technotexts, and Theories* (MIT Press, 2006).

Ghislain Thibault is an assistant professor in the Department of Communication Studies at Wilfrid Laurier University (Canada). His research and teaching include media theory, media archaeology and digital media studies. His work has appeared in journals such as *Configurations, the Canadian Journal of Communication and Intermédialités*.

Laura Watts is an ethnographer and associate professor at IT University of Copenhagen. Her research is concerned with the effect of landscape and location on how futures are imagined and made, and with ethnographic methods for writing futures. She has worked with the mobile telecoms industry, the renewable energy industry, and the public transport sector, and is currently working on the project *Alien Energy*, a collaboration with the people and places of the marine energy industry, particularly in Orkney, Scotland. She is co-author of the artisan book *Data Stories*, a set of ethnographic stories that engage with ideas of "big data." Much of her work is published on her website at www.sand14.com.

J. Macgregor Wise is a professor of Communication at Arizona State University's West Campus. He is the author of *Exploring Technology and Social Space, Culture and Technology: A Primer* (with Jennifer Daryl Slack), *Cultural Globalization: A User's Guide*, and the second edition of *MediaMaking: Mass Media in a Popular Culture* (with Lawrence Grossberg, Ellen Wartella and D. Charles Whitney). He most recently edited *New Visualities, New Technologies: The New Ecstasy of Communication* (with Hille Koskela).

Index

Page numbers in italic format indicate figures and tables.

active aging 188, 192
advertising/advertisement: focus on
    affective qualities for 124;
    for mobile Internet *141*;
    revenue generated by 71; for
    smartphones *142*; *see also*
    marketing
affect, movement and 235–9
affluent seniors 195–6
affordable housing 208, 210
*After Method: Mess in Social Science
    Research* (Law) 3, 202
aging population: active aging and 188,
    192; interviews with seniors and
    192–8; media studies and 187;
    mobility and 189–91; silencing
    of 189–91
Alliance for Wireless Power
    97, 102
alternating current system 92, 96
"always on, always connected" slogan
    117, 123
"always on" prosumer 112–13
Android operating system: Internet
    openness and 39; mobile
    handsets and 137–8; mobile
    Internet and 142; producing
    apps for 35
app developers: labor unions by 68;
    terms and conditions for 35;
    working lives of 67–8
Apple: advertising firms acquisition by
    71; app development process
    at 35; app ecosystems 235;
    wristphones 159; *see also* iPhone
    and iPad

Apple's App Store: apps availability and
    234–5; apps design and 67; top
    free apps in 240
apps: description of 233, 234;
    ecosystems 234, 235, 244, 245;
    fascination with 233–4, 238;
    individualist fantasies and
    240–4; individualization of
    consumers and 242–4; marketing
    of 239; widely popular 240; *see
    also* communicative capitalism
apps development: crowdsourcing of
    67; design issues 55, 66–8; for
    iOS and Android 35; labors of
    mobility and 66, 68, 69, 71;
    for mobile devices 34–5; open-
    source mobiles and 39–40; open
    standards issues 36
Arab Spring uprisings 138
assemblages: 4G Hot Spot and 204–10;
    of attention 214–15, 223;
    BlackBerry capitalism 111–13;
    Clickable World 213–14, 228;
    components in 203; concept
    of 202–3; description of 2;
    ethical obligations of 225–7; of
    mobile Internet 2–3, 6, 136–9;
    surveillance 225; theory of 203,
    210, 226
assembly: circuit of exploitation and
    64–6; e-waste trade and 73;
    workers 76
Association for Competitive Technology
    (ACT) 241
A Study of History (Toynbee) 90
attentional assemblage 214–15, 223

automobile-based transportation system 182–3

Being Digital (Negroponte) 97
BlackBerry: branding issues 116–18, 121; email through 33, 115; fanatical devotion to 113–14; fastest adopters of 121; as a lifestyle necessity 124; marketing campaign for 117–18, 121–2; POTUS story about 105; slogan 114, 117, 123
BlackBerry capitalism: assemblage of 111–13; fall of 125–6; intellectual property and 121; introduction to 105–6; materialities and imaginaries and 113–25; mobile Internet and 118, 121; understanding 106–10
BlackBerry Enterprise Server (BES) 119–20
Blade Runner (movie) 152, 162
body-to-screen: Lefebvre's windows and 218–23; screen in hand and 215–17; seeing through 217–18
branding issues 116–18, 121–2, 124
broadband Internet 137
broadcasting power 92–6

call center workers 71–3
camera phones 144
camera-to-object 223–4
Cannes, 3GSM World Congress in 155–6, 159–60
cell towers 48, 49, 56, 58
cellular telephones *see* mobile phones
circuit of exploitation: assembly and 64–6; design labor and 66–8; disassembly and 73–5; extraction and 62–4; introduction to 60–2; mobile work and 68–71; moments in 61; support and 71–3
Clickable World assemblage 213–14, 228
cloud computing 60, 100
coltan extraction 62, 63
communication: mobility and 22–3; practices 96, 98, 108, 109, 189; transportation and 19, 23
communication technologies: labors of mobility and 61; mobile information and 16; as taken-for granted technology 173

communicative capitalism: call center workers and 71–3; concept of 60–1; design labor and 66–7; digital workforce and 68, 71; e-waste trade and 73–5; introduction to 5; mobilizing imaginaries and 75–7; network exploitation and 239–40; reflexivity and 244–5; *see also* circuit of exploitation
communicative networks: Internet openness and 33; mobile access to 15, 17, 20
complex networks 240
concept of drive 236–7
consensus, politics and 18–19
Control Revolution, The (Beniger) 108
corporate investment in ICTs 107
critical mass: description of 171; diffusion and perception of 173–6; of mobile phones 184
crowdsourcing 67, 69, 70
culture, technology and 27–30

data collection and surveillance 138, 139
Democracy and Other Neoliberal Fantasies (Dean) 240
design labor, circuit of exploitation and 66–8
digital capitalism 106–8
digital connectivity 203, 204
digital information technologies 107, 108
digital media: digital economy and 100; features of 109; media practices and 109; personalization of 111; production costs and 110; ubiquitous connectivity and 111
digital networks 19, 100
digital piecework platform 69, 75, 240
digital technologies: aging population and 192; flexible employment and 20; homeless population and 205; mobile Internet and 136, 137; user-generated content through 144; working people and 18, 20
digital workforce 68, 71
disappearing technologies 226–8
disassembly, circuit of exploitation and 73–5
distracted consumption 216
domestication approach 174
drive as a reflexive force 236

electrical power 88, 92, 93
email: scanned documents and 175;
    through BlackBerry 33, 115;
    through mobile data service 32;
    wireless 117, 119, 121, 123
embodiment theory: locative media
    and 49; mobile media and 52;
    phenomenology of 53, 54
emergency shelters 209
Equinix Data Center: author's trip to
    45–6; introduction to 5; Peering
    Point 46, 47
etherealization, progress and 90–2, 96,
    98, 100
European App Economy 239, 241, 242
e-waste trade 73–5
extraction, circuit of exploitation and
    62–4
EZPass toll system 227

Facebook: as established social network
    system 138; free Wi-Fi by 245;
    homeless population and 205,
    206; locative media and 51;
    users' privacy issues 144
Fairphone company 76–7
fax machines: cost of 175; history of
    171; legitimation systems and
    176; postal workers strike due
    to 177
Federal Communications Commission
    (FCC) 33, 241
fiber-optic cables 47, 48, 49
file and content sharing 143–5
flexible employment 20, 67, 73
4G Hot Spot 204–10
4G mobile networks 1, 137, 159
Foursquare's databases 48, 51
Foxconn factory 64–6, 232
freedom: of movement 17, 19, 23, 90,
    100, 183; time and 20–1
free or open-source software (FLOSS)
    practices 31, 32, 40
FreeRunner 37, 39
Free Software Foundation 31
future archaeology, final thoughts on
    163–5

Geography of Time (Levine) 181
Gigwalk company 70
Google: advertising firms acquisition
    by 71; Android smartphone and
    39, 137, 142; app development
    process at 35; app ecosystems 235

Google Glass 217
Google Play 235
Google Wave 175
GPS coordinates 45, 49, 51

hand-tool relation 216
haptic and visual attention 213,
    216, 217
Hertzian waves 87, 90
homeless population: assemblages
    theory and 203; counting of
    208–9; description of 200;
    digital technologies and 205;
    Facebook use by 205, 206; gypsy
    mentality of 207; "Homeless
    Hotspots" event and 201–2;
    introduction to 200–1; Link-SF
    site for 208; mobile phones for
    206; mobilities paradigm
    and 206–7; slower movement
    of 207
homeless students 204, 209
hypermedia 217–18

identification, individualization of
    consumers and 244
imaginaries: in the age of ubiquity
    160–3; description of 134–5;
    of informational capitalism
    114; of the Internet 27, 29,
    32; materialities and 113–25;
    mobilizing 75–7; of wireless
    power 100
i-Mode system 136, 137, 140
individualist fantasies 240–4
individualization of consumers 242–4
informational capitalism: elements
    of 110; imaginary of 114;
    intellectual property and 110;
    introduction to 106; mobile
    Internet and 3; prosumption
    and 113; social mobilization
    and 112; theories of 109–10;
    ubiquitous connectivity and
    111–12
information communication and
    communication technologies
    (ICTs) 107, 112
information technology: culture
    and 27–30; digital capitalism
    and 108; migrant assembly
    workforce and 66;
    understanding diffusion of
    173–6; users as cocreators of

180; virtualization through 119; wireless power and 96, 98, 99
infrastructure issues: material 5, 22, 23, 90; mobile technologies 56, 57; wireless 201, 204
in-home refrigeration 183
Instagram 178, 240
institutionalization of social practice 181
integral accident 227, 228
intellectual property: ACT group and 241–2; BlackBerry capitalism and 121; informational capitalism and 110; mobile work and 68; modding hardware and 41; QWERTY keyboard as 116
International Electromagnetic Fields Project 98
Internet backbone 2, 201
Internet connectivity 5, 33, 200, 204
Internet Engineering Task Force (IETF) 30, 31
Internet Governance Forum (IGF) 30, 31
Internet Imaginaire, The (Flichy) 4, 139
"Internet of Things" 4, 138
Internet openness: background 26–33; communicative networks and 33; conclusion about 40–1; mobile enclosure and 33–40
interpersonal interactions: facilitating 171, 172; use of texting for 179
interviews with seniors 192–6
involuntary immobility 5, 16
iOS operating system 35, 235
iPhone and iPad: as affective machines 233; assembly of 64; hypermedia and 218; open standards issues 39; RIM's decline due to 125; *see also* mobile devices/media
isolation, individualization of consumers and 242–3
iWatch: introduction to 149–50; vs. Pebble watch 151–2; PEBL mobile phone and 154; Swatch Talk and 161

Jana company 69, 70

labor conditions 65, 74, 227
labors of mobility 61, 68, 75, 77
labor unions: by app developers 68; call center workers and 72, 73
labor unrest 66, 72

landline telephones 174, 177–9
Lefebvre's windows 218–20, 223
legitimation systems 176, 184
Link-SF site 208
Linux open-source operating system 36, 39
LiveCams apps 214, 218, *219*
location-aware technologies 49, 55
locative media: audience of 50–2; Equinix Data Center and 45–50; file and content sharing through 145; introduction to 45; mobile Internet and 138; object-oriented phenomenologies and 52–5

Marconi's experiments 91, 93, 94, 99
marketing: of apps 239; for BlackBerry 117–18, 120, 121, 123; focus on affective qualities for 124; social repercussions of 196–7
MasterCard PayPass system 227
material infrastructure 5, 22, 23, 90
materialist medium theory 108
materialities: of camera 223; imaginaries and 113–25; for mobile devices 33; of mobile Internet 3–4; traditions invoked by 3
mechanical refrigeration 183
mechanical timekeeping 181, 182, 183
media forms 109–11, 145
media materialism 3, 108
media practices 40, 109, 190, 235
media studies 187, 189, 190
mediation technology 174, 176, 183
media uses and practices 187, 200
medium theory 3, 108, 109
Metcalfe Law 173
micro-work system 69–70
migrant assembly workforce 65–6, 76
mining issues 62, 63
mobile applications development *see* apps development
mobile apps *see* apps
mobile assemblage 188, 192, 196
mobile broadband 137, 141, 142
mobile communications: being always available through 120; connectivity issues 2, 5; digital capitalism and 107; early stages of 1; material forms of 3–4; texting as a part of 178; *see also* circuit of exploitation
mobile data 54, 136, 137, 244

mobile devices/media: apps
development for 34–5; branding
issues 124; consumer choice and
34–5; design issues 55; hand and
screen relationship and 216–17;
hand-me-down 196, 197;
increasing use of 32; marketing
campaign by 196–7; materialities
for 33; Openmoko project and
35–40; phenomenology of 52–5,
216; standards and protocols
33–4; as a survival device 205;
visibility and invisibility issues
55–8
mobile handsets 6, 137, 139
mobile infrastructure 56–8, 159
*Mobile Interface Theory: Embodied
Space and Locative Media*
(Farman) 52
mobile Internet: advertisement for
141; assemblages of 2–3, 6,
136–9; Blackberry capitalism
and 118, 121; changes brought
by 204; conclusion about
145–6; description of 1; digital
technologies and 136, 137;
file and content sharing and
143–5; imaginaries of 4, 135–6,
140–1; imagining 139–43;
intense development and
activity for 137–8; introduction
to 1; locative media and 138;
materialities of 3–4; multiple
temporalities 5; platform OS and
118; prominent representations
of 141–3; shift from wired
Internet to 25; social networks
through 138; valorization of 15,
16, 17
mobile Internet architecture: conclusion
about 40–1; consumer choice
and 34–5; introduction to 25–6;
Openmoko project and 35–8;
openness issues 26–7, 33–4;
open-sourcing and 38–40;
shift in cultural norms and
32–3; technology and culture
and 27–30; wired Internet and
30–2
mobile media *see* mobile devices/media
mobile networks/technologies: aging
population and 189–90, 196–8;
infrastructure issues 56, 57; for
mobile broadband 137

mobile phones: adoption rate 187;
branding issues 124; critical
mass of 184; diffusion of 175;
hand-me-down 196, 197; hand-
tool relation 216; for homeless
population 206; locating spouse
through 194–5; modding of 41;
pebble-shaped 152–3; roaming
issues 195–6; service contract
193, 195, 196; SIM-less 193–4,
197; SMS texting and 180;
statistics about 235; *see also*
mobile devices/media
mobile phone users 32, 178, 235, 240
mobile populations *see* homeless
population
mobile practices: busing at night 193;
safety issues 192–3; scooting
around 193–4; spousal locator
194–5; stories-so-far concept
191–2; Wheel-Trans service 194
mobile screens: articulations and
infoldings and 224–8; aspects
of 214–15; body-to-screen and
215–23; camera-to-object and
223–4; introduction to 212–14;
phenomenology of 215, 216;
seeing through 217–18; swarm
of 213
mobile technologies *see* mobile
networks/technologies
mobile telecoms: final thoughts
about 163–5; imaginaries
and 160–3; industry vision
157–8; innovation issues 154;
introduction to 149–50; pebble
futures and 151–5; speed of
change in 156, 157; ubiquitous
futures and 155–8; un-dead
futures and 150–1, 158–60
mobile television 138, 216
mobile work, circuit of exploitation and
68–71
mobile workforce 112, 120
mobilities paradigm: communication
and transportation and 19;
homeless population and 202,
206–7; introduction to 2–3
mobility: aging population and 189–91,
196–8; communication and
22–3; emergency purpose
for enhancing 195; freedom
and 17, 19, 23, 90, 100,
183; homelessness and

205; transportation and communication and 19–20; valorization of 17; work can/cannot wait issue and 20–1
Moore's Law 73, 75, 155, 157
Motorola company 67, 70, 149, 153
mundane affairs, coordinating 171, 172

National Alliance to End Homelessness 200, 205
Neo device 36, 37
network exploitation 239–40
network information economy 100, 109
network neutrality 32, 33, 34
New Spirit of Capitalism, The (Boltanski and Chiapello) 17
Nintendo DS (dual screen) mobile gaming system 217
Nokia company 65–7, 136, 140
nongovernmental organizations (NGOs) 27, 29

Obama phones 206
object-oriented phenomenologies 52–5
object permanence concept 227
online labor brokers 67–8
On the Move (Cresswell) 16
opaque media 218
open architecture: introduction to 5; results of 39–40; software development through 26, 36
Open Handset Alliance (OHA) 39
open Internet *see* Internet openness
Openmoko phones case 35–41
open networks 29
open-source software development 26, 36, 38, 40
operating system *see specific types*
overhead wires and power lines 88, 89

packet-switched data service/networks 114, 118, 136, 137
PDA devices 115, 117, 188, 212
pebble-shaped mobile phone concept 152–3, 155
Pebble watch 153
peer governance 30–1, 35, 39
peer-to-peer (p2p) networks 144, 145
personalization, individualization of consumers and 243
phenomenology: of embodiment theory 53, 54; of media devices 216;

of mobile screens 215, 216; of sensory inscription 54
platform OS 118
political imaginary 16, 17
politics: consensus and 18–19; defined 15, 19; mobility and technologies and 19
politics of immobility: introduction to 5, 15–16; mobility valorization and 16, 17; *see also* freedom
politics of mobility 5, 15, 16, 200
postal workers strike 177
Power Matter Alliance 97, 102
prosumption 112–13
push-based communication 114, 117, 118

QWERTY keyboard 115, 116

radio broadcasting 95, 100
raw materials extraction 62–4
reflexive expectations 176–7
reflexivity of the app ecosystem 244–5
refrigeration technology 183
Research in Motion (RIM): annual report of 123–4; branding issues for 116–17, 121–2; decline of 125–6; enterprise server by 119–20; introduction to 105
responsibilization, individualization of consumers and 243–4
roaming issues 195–6
Roger's model of diffusion 173

safety and security issues 171, 192–3, 196, 197
Samsung 67, 97, 149
science fiction futures 163
scooting around 193–4, 198
seniors *see* aging population
sensory-inscribed phenomenology 54
Silicon Valley Toxics Coalition 74
SIM-less cell phones 193–4, 197
situated futures 158–9
smartphone cybertariat 61, 75, 76, 77
smartphones: advertisement for 142; Internet access through 138; open-source 36, 39, 40; *see also* circuit of exploitation; mobile devices/media
smartphone users *see* mobile phone users
SMS texting 1, 172, 178, 189

sneaky seniors 195–7
social constructionism 179–81
social ecology, changing 177–8
social imaginaries: Clickable World 213; description of 4; Internet openness and 27; of mobile Internet 135–6, 140–1; social practices and 4, 77, 201; wireless telegraphy and 92
social interactions: coordinating 171, 172, 180; timekeeping for 182
social media: advertising revenue from 71; circuit of exploitation and 60, 66; prosumption and 112–13
social mobilization 111, 112
social networks: BlackBerry smartphone and 124; through mobile Internet 138; ubiquitous connectivity and 114; Web 2.0 prosumers and 123
social practices: institutionalization of 181; of open governance 40; social imaginaries and 4, 77, 201
software development 26, 36, 38, 40
software development kit licenses (SDKs) 35, 39, 40
Sony Ericsson 64, 67, 136
Special Economic Zones (SEZs) 64, 65
spousal locator feature 194–5
stories-so-far concept 191–2, 196–8
stupid network idea 29
suburban housing 181, 182, 184
support, circuit of exploitation and 71–3
surveillance assemblage 225
surveillance camera 223, 224
survival devices 205
Swatch Talk 149, 161, 162

tablets *see* mobile devices/media
tactile vision 216, 217
taken-for-granted technologies: conclusion about 183–4; elements of 173–8; introduction to 171–3; other examples of 181–3; social constructionism and 179–81; texting as a 178–9
Tarnac Nine case 21, 22
TCP/IP protocol 34, 136
technological assemblage 215
technological imaginary 6, 140, 162
technologies of visual attention 226–8
technology *see* information technology

telecommunication network 75, 107, 158
Telecommunications Act of 1996 56
telework 112, 118, 119
Tesla, Nikola 93–4, 96, 100–1
text messages/messaging 69, 137, 156, 172, 178, 183
Thames Valley region 155, 159
3G networks 1, 47, 137
3GSM World Congress 155–6
thumb culture 115, 116
timekeeping system 181, 182
Touch Watch Phone 149
tourist and traffic cameras 223
transparent media 217
transportation technology 19, 23
Tubes: A Journey to the Center of the Internet (Blum) 45
Twitter 51, 201, 205
2G networks 134, 137

ubiquitous connectivity 111–12, 122–3, 125
ubiquitous futures 155–8
un-dead futures: fieldsite for 150–1; landscapes of 158–60
Union Network International 77
unmanned warfare 95
Unwired Planet 136
U.S. Department of Housing and Urban Development 200
user-generated content 144

Viber app 172, 178
virtualization of capital 112
virtual organization 118–20
virtual workforce 112
visual attention technologies 226–8

Web apps *see* apps
webcam applications 216, 218, 220–2
Web 2.0: always on, always connected concept and 123; informational capitalism and 110; prosumption and 113; *see also* mobile Internet
well-off seniors 195–6
Whatsapp 172, 178, 240
Wheel-Trans service 194
Wi-Fi technology 1, 136, 137, 245
wired Internet 25, 26, 30–5
wired technologies 88–90, 98
Wireless Application Protocol (WAP) 136, 137

wireless bandwidth 1, 4
wireless charging technologies 97
wireless data networks/technologies
112, 118
wireless email 117, 119, 121, 123
wireless industry *see* mobile telecoms
wireless infrastructure 201, 204
wireless Internet: connectivity
issues 204; description of 1;
introduction to 5, 6; Wi-Fi
technology and 150
wireless power: arguments against
97–8; conclusion about
99–100; current state of 96–7;
electromagnetic hypersensitivity
and 98; ether and 97; features
of 94–6; imaginaries of 100;
information technology and
96, 98, 99; introduction to
87–8; main application of 98–9;
Marconi and Tesla projects and
92–4; wired technologies and
88–90
Wireless Power Consortium 97, 102
wireless telegraphy 87, 90–3, 134
WiTricity 97, 99
working people, digital technologies
and 18, 20
World Summits on the Information
Society (WSIS) 31
World Wireless System 161
wristphones 149, 150, 155, 159

zombie futures 160, 162, 164, 165

Milton Keynes UK
Ingram Content Group UK Ltd.
UKHW031533071024
449327UK00005B/70

9 780415 731003